WITHDRAWN

Sight and Sensibility

The Ecopsychology of Perception

Sight and Sensibility

Laura Sewall

Jeremy P. Tarcher / Putnam · a member of Penguin Putnam Inc. New York

Most Tarcher/Putnam books are available at special quantity
discounts for bulk purchases for sales promotions, premiums,
fund-raising, and educational needs. Special books or book excerpts also
can be created to fit specific needs. For details, write Putnam Special
Markets, 375 Hudson Street, New York, NY 10014.

Picture of Closeup of Nautilus Shell Spiral by CORBIS/Bettmann
Picture of Nevada Desert Dunes by CORBIS/Chris Rainier
Picture of Calladium Leaves by CORBIS/Karen Tweedy-Holmes
Picture of Looking up Into an Aspen Grove by CORBIS/Darrell Gulin

Jeremy P. Tarcher/Putnam
a member of
Penguin Putnam Inc.
375 Hudson Street
New York, NY 10014
www.penguinputnam.com

Library of Congress Cataloging-in-Publication Data

Sewall, Laura, date.
    Sight and sensibility : the ecopsychology of perception /
Laura Sewall.
        p.   cm.
    ISBN 0-87477-989-8
    1. Apperception.   2. Perception.   3. Awareness.   I. Title.
BF321.S45   1999   99-28702 CIP
152.1—dc21

Printed in the United States of America
10   9   8   7   6   5   4   3   2   1

This book is printed on acid-free paper. ∞

Cover art: *The Channel*, painting by Joan Solomon. For information on
The Art of Meditation: www.joansolomon.com; (612) 926-5125.

Book design by Chris Welch

This book is dedicated to my mother, Phebe Buckman Sewall,
who showed me very small shells in a handful of sand,
my first memory and my first lesson in learning to see.

Acknowledgments

It is impossible to know where to begin in giving my thanks. As I turn toward the last year of writing, I see many vivid images of the friends and family who have sustained me in the process. There is Liz Faller calling me to the highest task and loving me all at once. There is Julie Munsell generously listening to my theories while riding a mountain bike uphill, the sun and Ponderosa forests illuminating her. I see Kevin Keith appearing as if dropping from the sky, a Sky God with laughter and the wisdom to carry me away from the computer. There is David Abram encouraging me when I was least certain, Chellis Glendinning singing a victory chant to me on the phone, and my wise friend Peggy Bill reminding me that friendship is more important than any book. I see Wayne Regina, the dean at Prescott College, whose belief in me—and in the importance of a thoughtful book on perception—made this project possible. And I have many, many thanks to give to my vibrant students, who continuously offer me so much inspiration.

I see my father, Bill Sewall, lonely after my mother's death, kindly and patiently waiting for me to have time to talk. I hear my sister, Abbie, admonishing me—with love—to "just finish the book," and my sister, Gretchen, offering wise counsel. My third sister, Lee, is laughing, trusting that I am simply following a God-given path. I see Bill Plotkin, who has ever so gracefully and patiently listened and waited, believing in the process of doing one's soul work, of my need to write. Without his faith, I would not have been able to complete this book with such love.

Thank you to James Hillman and Ted Roszak for their brilliant writings, insight and guidance. I give my thanks to Walt Anderson, a naturalist who has an eye and graciousness like no one else I know, and to Glendon Brunk, who within several moments of coming to know me, reminded me that the bottom line is love. My thanks to Robert Greenway for his truly crazy wisdom and friendship, and to Margaret Clark, who always believed that I would be writing this book. I am grateful for the many conversations and gestures of encouragement from Nina Utne, Liza Bakewell, Tom Fleischner, Grace Burford, Steve Munsell, Michelle Apland, Laragh Kavanaugh, and Pat Trott. And thanks to Ellery Kimball, the wild young illustrator who loves life with every cell of her body; and to Anne Depue, my ever so generous agent; to Joy Parker, my insightful editor and true comrade, and to Wendy Hubbert, my editor at Tarcher, whose intuition made this project possible.

Endless, these thanks. My thanks are for the ponderosa pines and Arizona oaks that stand just outside the windows where I have written for every possible hour over the last year. And for the tides I contemplated on the coast of Maine, their ceaseless rhythm both pressing in on me and pulling me back into the world again. Thanks for the songs, and the friends that sang to me, for David Gilligan singing in the last moments of writing:

Oh, I see spirits,
and they dance around me . . .
Say the birds and the mountains,
they all speak to me . . .
saying "come along, come along with me,
come along, come along, come along with me. . . ."

I give my greatest gratitude to the truly divine spirits that keep coming to me—the old growth trees as they fall, the elephants I have found slaughtered for ivory, the pure waters as they slip away, the special glow of a storm gathering. All of this experience is about being in the world at this time. I am thankful for the times. They are calling us into greater consciousness, into a renewed and deepened sense of both spirit and soul.

My final thanks go to the world itself, our precious planet.

Contents

The hour is striking so close above me,

so clear and sharp,

That all my senses ring with it.

I feel it now; there's a power in me

to grasp and give shape to the world.

I know nothing has ever been real

without my beholding it.

All becoming has needed me.

My looking ripens things

and they come toward me to meet and be met.

No thing is too small for me to cherish

and paint in gold, as if it were an icon

that could bless us

though I'll not know who among us

will feel this blessing.

—Rainer Maria Rilke

Foreword

by David Abram

Vision—the sense of sight—has been roundly disparaged in recent years, particularly among those who are dismayed by the destructive legacy of modernity. Mounting social injustices and intensifying disarray within the larger natural world have led many to question the grounding assumptions and practices associated with contemporary civilization. This has resulted in a curious indictment of the visual sense among many cultural theorists. Such thinkers tell us that vision, more than any other sense, has lent its particular character to Western civilization, and indeed that the cool detachment and objectification associated with the modern, technological enterprise can be understood as the result of an overemphasis on visual ways of interacting with the world. Vision, they claim, is the most distancing of the senses—after all, we must step back from anything in order to *see* it—and alone among the senses, vision invites us to *objectify* the world, to coolly observe the things and events around us rather than to participate in those events. In contrast to touching, tasting, or even hearing, an over-engagement in *seeing*—according to these theorists—is at least partly to blame for our contemporary recklessness toward the earth and toward one another.

But is this disparagement of vision really warranted? To be sure, the unique character of European civilization is partly a result of the early influence, in the West, of alphabetic writing and "The Book"—and it is possible that the growing dependence upon visual texts precipitated a gradual diminishment of the other senses, a blunting of auditory and tactile modes of relationship. Nevertheless, while literacy may have granted a new centrality to the visual sense, it was not really vision, per se, that was elevated, but only a very limited *kind* of vision—a close and narrowly focused gaze that lent itself readily to instrumental modes of engagement.

"The objectifying gaze" has become a cliché of contemporary criticism. But what of the eye of the heart? What of the contemplative gaze, or the stare of astonishment? What of the appreciative eye, or the expectant gaze, or the glare of rage? What of the eye of the visionary, or the lover's gaze? Surely we need, today, to be introduced afresh to the fluid nature of seeing—a new aquaintance with the manifold powers of sight. And that is precisely what Laura Sewall sets out to provide in these pages: an invitation to genuinely *see* in new ways.

Sewall asks us to take seriously the ancient and medieval notion of an inner light, an ocular fire that emanates from the eye, illuminating the things seen. She writes of seeing, in this sense, as an *act*, a dynamic "raying out" of vision into the world. Yet she also writes of being penetrated by the visible:

> [W]e ingest billions of photons during every open-eyed moment. As waves, light vibrates into us. It is transduced, within

several layers of retinal tissue, into electro-bio-chemical pulses surging deeply into the tissue within the mammalian brain. It shifts from wavelength to color, from photons to pulse—and from pulses to hormonal flushes. . . .

She writes of being invaded and possessed by the shifting colors of grasses, satiated by moonlight, and awash with the tides swelling the marsh where she lingers. She speaks of being permeated by the visible.

Oscillating between the experience of vision as a fiery emanation raying out from the seer, and the contrary experience of sight as an influx or invasion by that which one sees, Sewall's exploration gradually gives rise to a new sense of vision as a deeply reciprocal event, a participatory activity in which both the seer and the seen are dynamic players. To see is to interact with the visible, to act and be acted upon. It is to participate in the ongoing evolution of the visible world.

Nothing could be farther from conventional assumptions than this recognition of vision as a reciprocal or participatory interplay between the animal eye and the animate cosmos. Such a recognition confounds the subject/object epistemology that undergirds so much of the modern world—the notion of the self as a pure subject gazing out at a natural world construed as a basically determinate set of objects. This dichotomous epistemology entails that our relation to the earthly world is that of spectators looking at a kind of spectacle; it ensures that there is no real reciprocity between the perceiver and the perceived, no real affinity between humankind and the rest of the earth. Nature has thus become, for most moderns, little more than a pretty backdrop for a picnic, or something to stare at in the zoo.

No doubt this spectator-epistemology has been fostered, in the West, by our increasing propensity to stare at flat surfaces—by our habit of focusing our eyes, steadily, upon the flat pages of books and magazines, and by staring, hour after hour, at the flat screen of our television or our computer. The remarkable nature programs that we watch on television may teach us fascinating things about the mating habits of giraffes and the migratory flight of geese, but what they really communicate to our sensing bodies is that nature is something we look *at*, not something that we are *in* and *of*. By ceaselessly training our gaze upon these flat surfaces, we rob our eyes of their regular engagement with the shifting depths of a world that unfolds from the near to the far, from the tips of our fingers to the far-off horizon.

While a flat screen is something we gaze *at*, a world that exists in depth is something we gaze *into*, continually shifting our focus from the spider weaving its web just in front of our nose to the distant clouds massing on the horizon, and from there back to the swallow swooping over the neighbor's field. Depth—the dimension of closeness and distance—exists only when we are entirely included *within* the landscape that we perceive, such that as we shift our position, the landscape's depth shifts around us (a river valley suddenly opening up before us as we crest the hill, while the snow-capped mountain that had been drawing our gaze quietly vanishes behind a group of cedar trees). If we no longer consider ourselves to be wholly a part of the natural world, it is perhaps because we have lost our *depth perception*, because our eyes have forfeited their native ability to continually respond to the close and distant beckonings of an enveloping terrain—because, that is, we have become accustomed to looking *at* the world rather than gazing out into its depths. It is for this reason that

Sewall strives, in this work, to reawaken our perception of depth, this most enigmatic and mysterious aspect of the visual field.

The detractors of vision know nothing of this fluidity, the way our eyes, when truly awake, open us onto a world of wonders. In truth, the realm opened to us by our eyes is not merely the visual field, but the whole of the sensory cosmos. For vision, it turns out, simply cannot be understood in isolation from the other senses. The eyes are not autonomous organs, and the way we see things is profoundly influenced by what we hear or even taste of those things, by the way we imagine their textures would feel to our fingers or against our skin. Indeed, vision may well be the most *synaesthetic* of the senses—the sense most thoroughly infiltrated and altered by the participation of the other senses. Just as our fingers can see for us in the dark, so we can *caress* surfaces with our gaze—we can explore the ruffled textures of tree bark and lichen by the way we move our eyes across them. Similarly our eyes, when really awake, can taste the sweetness of ripe blackberries glistening among the thorns even before we pluck them. Meanwhile, the participation of our ears lends a kind of receptivity to our gaze: We find ourselves listening with our eyes, and even hearing things, too—as when we move our visual focus across these printed letters. Cézanne felt sure that if he painted a still-life just right, using just the right mix of colors and applying them to the canvas in just the right manner, then even the *smell* of those apples would be present in the painting!

In order to think deeply about vision, then, we must learn to think not only with our eyes but with our entire body. Such is the vitally important endeavor that Laura Sewall urges us toward with this book. She is clearly a woman who thinks not only with her brain, but with her senses and with her heart. Let us now see what she has discovered.

Introduction

If I were to wish for anything,
I should not wish for wealth and power,
but for the passionate sense of the potential,
for the eye which, ever young and ardent, sees the possible.

—*Søren Kierkegaard*

There was a time when I never thought about seeing. Seeing came quite naturally, and like the primal peoples who never needed a word for nature, I had no thought about the fact of having vision. But by the time I was nineteen I could not see the stars my friends kept arranging into constellations and stories. My vision had changed after several months of living on an anxious edge in Kenya—eight miles down a dirt road to Uganda and eighteen miles in the other direction to Kakamega, the nearest town. I taught algebra to sixty girls, some older than myself, went to the market in Kakamega on Saturdays, and entertained the children who, never having seen a white woman, stood for hours outside my window, staring and giggling at my odd ways. I was frightened by my white-girl innocence, by the voodoo that surfaced during local elections, by the embezzlement I witnessed and reported, and by the subsequent warnings to track everything I owned in order to prevent any black magic cast my way. As a decent defense, I neurotically picked at the split ends of my hair, thinking and thinking and hiding in the self-contained world of my psyche. In the process, I quickly became quite nearsighted. Still, I never thought about seeing, at least not until I returned to the States and filled my first prescription for glasses.

I began to wonder more about seeing, and not-seeing, over the next few years while kayaking off the west coast of Vancouver Island, in Barkley Sound, Nootka Sound, and Esperanza Inlet. My glasses were constantly crusty with salt spray. I took them off to see possible campsites, squinted, couldn't see, cleaned the lenses, put them back on, and still couldn't quite make out the world. But one thing did become clear—there was much more to see than what I had presumed to be good enough. And so I packed up my red truck and country girl ways to study the Bates Method of vision improvement with the best teacher I could find. Unfortunately, she was in Los Angeles. I drove steadily south along Highway 101 until I reached the outskirts of the city. Terrified of cars and traffic and people, I hovered on the edge of the city until late at night when, I presumed, the traffic was minimal. Even then I held my breath, clinging to the steering wheel and peering over the top edge, as if a 20/20 correction wasn't enough to help me find my way.

I began my lessons. I did whatever my teacher, Janet Goodrich, asked me to do. I took off my glasses as often as I could, covered my eyes, and visualized velvet blackness. I imagined quiet forests and restful beach scenes with wind blowing through tall palm trees. In my imagination, the sound of wind and waves soothed me, the ocean was my favorite color of turquoise, and the edges of all I imagined were crystal clear. I did attention exercises, eye-

balancing exercises, and "near-far swings." I made a habit of looking into the distance, then back to the point where my vision
cleared, then breathing and stretching my clear vision into the distance again. My eyes eventually learned to rest a bit when I looked
into the distance, less fearful of not seeing clearly and more comfortable with whatever was there. The landscape became more
distinct, more apparent—or was it my imagination? I continued
my exercises, rubbing my palms together in earnest now, covering
my eyes with warmed hands several times a day, visualizing places
of beauty and serenity and resting the muscles around my eyes.[1]
My imagination became strong and clear and I was able to picture
whatever I wanted to see. The world began to light up with a newfound resonance and my resistance to the urban landscape, to Los
Angeles and to the world in general, diminished. I became more
receptive to seeing the whole of what lay before my eyes. And so
I discovered that the lessons included relinquishing my fears and
defenses and looking directly at what was keeping me from fully
facing the world. After a month or so, I suddenly glimpsed sharp,
razor-like edges and neon colors, as if the cosmos were simultaneously teasing me and gifting me with a miracle. But I could not believe my eyes and, sure enough, the crystal-clear world slipped
away. With time, however, more glimpses came, now teasing me
like insights slipping in and out of view. As I became more practiced, as I developed a kind of crafted nonchalance about this
great new beckoning of the world, glimpses became long moments of rich perception. Soon fabulous shapes and vibrant colors
signaled to me, edges were sharp all the time, and whole stories revealed themselves on street corners. I traveled all over the city
with a kind of gawky yet graceful wonder. Fear slipped into a previous life and even my troubled relationship with time became

finely tuned. I arrived on time everywhere and every time. I moved, it seemed, through space and time with a soft, synchronized rhythm, as if seeing extra clues for the first time.

So begins my story of learning to see. With more practice my acuity sharpened. The world came comfortably close to me, as if gently nudging and then pressing against me. Colors glittered and the world became tasty. Whispers of air slipped and wrapped around my eyes, my neck, and my ankles. Vibrations were unmistakable, as if speaking directly to me, and I got the messages. Within a shiny, shimmering new world, I became part of something big going on, vibrant and always calling for my attention.

Although my story is personal, like the feminist adage, "the personal is the political." While growing up I had never been given any indication that I had the power to change the way I saw the world or that such changes had radical implications for the quality of my life. For the first time, I was experiencing the power of my body—of my senses and my intention—and I became hungry for more. I was falling in love with the world and desperately curious about how my new vision was stirring up such a potent mix of sensations and emotions within me. And so my new view included a glimpse of the way our culture had disavowed the power of direct experience and, with it, the generous realm of the sensuous. I had to know more. I went to school to study vision, our potential to see. Being adventurous, my studies soon became research in the backcountry of East Africa.

Dr. Max Snodderly, of the Eye Research Institute in Boston, sent me off to the savannas of central Tanzania as a research assistant. My assignment was to collect color samples of the fruits eaten by yellow baboons. For most of a year, I followed several troops of baboons across wide stretches of savanna. Each day I wandered in the midst of great herds of wildebeest, gazelle, cape buffalo, and zebra; families of giraffe, lions, and elephants; packs of hyenas and wild dogs. Despite the exotica that surrounded me, I remember the long grasses best. They filled every glance with wide expanses of color, color that shifted from every shade of green to deep mauve as the rainy season passed into dryness.

Although I was there to study the visual behavior of baboons, I knew this was an arrogant enterprise the minute I witnessed the baboons' alarm call for leopards. I never saw the leopard, but they always did. This happened many times, and each time I immediately turned to run with the entire troop, not questioning their ability to see much better than I. Despite this clear indication of our relative abilities to see, I was to study theirs. In particular, I was to study their ability to discriminate color by recording and photographing the fruits they ate. The hypothesis was that the baboons' peak sensitivities to color, as measured in the lab, would match the particular wavelengths reflected off the fruits they chose to eat. There was a lot of jargon to wade through—nanometers, reflectance spectra, and spectral sensitivity curves—but the hypothesis was exceedingly simple, too simple, in fact. A kind of coevolution between the color of the most desired fruits and the sensitivity of the eye designed to see them seemed, from my perspective, an obvious fact of God.

I was far more interested in whether or not the baboons who

served as sentinels, sitting high atop tall tree stumps, had particularly good vision for distance, and if those baboons who seemed to do all the near-point grooming, for long devoted hours, were nearsighted. Without being conscious of the fact, I was hypothesizing about my own experience at Namirama Girl's School ten years before, picking at the ends of my hair and becoming myopic. More important, I wanted to know if long-distance looking meant long-distance seeing. I wanted to know if there was a direct relationship between the "practice" of looking into the distance, the nearly continual experience of it, and a heightened capacity to see way out across the savanna. This was a simple hypothesis but one that, surprisingly, was argued about a great deal in the literature on vision. Most everyone seemed to think that nearsightedness was inherited, that normal was satisfactorily defined by a Snellen eye chart as 20/20 vision, and that normal vision was equivalent to our perceptual potential. But I knew Dr. Snellen's story: He had established the standard for acuity by calling his assistant, who apparently had good vision, to the chart one day to measure what he could see from 20 feet away. And so it was that normal became normal. In any case, I wondered what was possible.

Wandering around in the savanna each day did little for the experimental method. There were far too many variables, such as the spiritual crisis my research partner was going through, a chaotic relationship, and the fact of numerous influences—of which we had no clue—on the baboons. Besides, doing research on the evolution of baboon color perception as a function of the colors in their landscape was, in my mind, redundant and ridiculous, although I knew it was funded to the tune of several hundreds of thousands of dollars.

I became much more interested in my experience than in Dr. Snodderly's experiment. The difference was in participation, and I couldn't help myself. I was in love with the long-stemmed grasses. In the early morning they glistened and softened my eyes. In the late afternoon they gathered and reflected low warm light, softening my body. They hid all sorts of exotic animals and grew to be so tall that wading through them was more like pushing against soft seven-foot walls. They were the ground of my entire world, filling each day with brilliant shades of chartreuse and deep green, then shifting with the season to peach tones, to salmon and pink; then to many shades of rose, burgundy, mauve, purple, and plum; then to yellow and shiny golden tones. All day long, every day, they gathered up my attention, swaying and teasing me, tugging me into a widening gyre of color, and of the world.

I had been in the field for seven or eight months when my love affair was consummated. I had just finished lunch and was sitting on a log looking west over several thousand acres of savanna. Charles, my Tanzanian companion, was somewhere off in the distance, near the front of the baboon troop. The troop was moving in a slow sweep across the savanna, spread over a mile or two in an area still moist with the last of the rainy season, where they quietly foraged for buried sedges. I sat at the tail end of the troop and, like most other days in the field, I could have been the only human on the planet. Gazelles and wildebeest grazed in huge herds on the horizon, with a long line of blue-hued mountains as backdrop. I remember shifting my gaze to the grasses and noticing the color changes from the day before—and then it came. I could say I was consumed, washed over by waves of *rightness*, is-ness and nothing more, but these words fall short of my experience. I had no doubt, no question, no thought until after the fact. I simply felt *yes, entirely this, totally as it should be,* without words or the thoughts to think them, bliss, rapture, unity. Sudden, swift, easy, like sway.

But like flashes of clear sight, the thought of the moment chased away the experience, and bliss slid into wonder. In the aftermath, I knew that light had streamed through my eyes, that I had been gathering light for many months, and that beauty had become deeply embedded in my brain, sinking into my body and resurfacing as bliss. This, I'm certain, is what mystics call the unitive experience; no doubt, I was unified. My preparation for the moment had been the many months of wandering in an unbroken landscape, silent and solo. I was alert, with every glance awakened and attentive while walking in the company of lions and cape buffalo, with pythons and mambas somewhere in the grass and elephants suddenly appearing out of ravines thick with bushes, shrubs, and trees. My eyes continually stretched to and rested on the distant blue horizon,[2] continually scanning, looking and watching, as if busy in the act of gathering beauty—although it seemed to seep through every pore and sometimes simply slice through me. I had surrendered myself. It was a naive offering on my part, and it was a great gift received, and given back, received and given back again.

Within the context of Western psychology, the unitive experience is both exotic and problematic and sometimes it is even considered pathological—but this description doesn't come close to the experience. The truth is, we are not sure of what to make of it. Some consider it presumptuous to make claims on the experience of enlightenment—but clearly, I was filled with light. It is also difficult to talk about such soul-satiating moments in a culture gone mad with machines. Such intangible moments just aren't real enough, like the metal and mass that surround us. And it doesn't help that we are buried in a language that is more interested in nouns and things than verbs and experience. Still, there are many names for the experience of enlightenment, suggesting that it is not uncommon or unknown. It is called *sahaj samadhi*, Nirvana, the "Gateless Gate," and the "One Taste." From my perspective, enlightenment is the unitive experience. It is our communion with the Other, the door through dualism.

We are the sum of our experiences. Jolt by jolt, I was learning to perceive beyond unspoken limitations, the canon of my day. Without awareness of the fact, I was developing a true vision, one that continues to inform and orient my life, one that still tugs at my senses and turns my gaze toward the light. I was coming to see that we are embedded in light and beauty, that we are receptacles, vessels, and transformers of the vast forces of the universe. But I had a deferred admission, a place waiting for me, in the psychology department at Brown University—purportedly one of the best places to study vision in the country—and my newfound mystical inclination could go no further. So I traveled on to graduate school to study how we see, or how we might possibly see. I found myself in the very heart of traditionalism with a very nontraditional experience of ecstasy—conveyed by light streaming through my eyes—provoking my research agenda, my undercover studies of visual potential. As I entered the academy, however, I carefully toned down my vision, limiting my agenda to simply seeing beyond the Snellen eye chart. To my professors, I never spoke about ecstasy or transcendent moments of seamless experience.

I wanted to learn everything about the visual process. The department came to expect stellar performances from me, with lots of data and many publications. But research methodology or, more specifically, reductionism narrowed my field and I nearly lost my vision, both metaphysical and physical, along the way. Without a vision of perceptual potential, I soon became much less than the star student. Furthermore, within three months I was wearing glasses again—the obvious and inevitable result of looking at black-and-white, two-dimensional text, eighteen inches from my eyes, for many hours each and every day. This, too, became a lesson in vision, perhaps the most substantive at the time.

Despite the loss of my clear and inspired vision, I continued my research. I read between the lines, asked many questions in carefully controlled labs, and pieced together a developing story: The neural structure of the visual system changes when the attentional processes in the brain are activated. It was assumed by the researchers that this was purely a developmental phenomenon, or that such change in the visual system occurred only during particular developmental stages in kittens and young monkeys, and that adult animals would not show notable structural changes. Among visual scientists the discussion of such fundamental changes in the visual capacity of adult animals was apparently heretical. Questions posed in research seminars about structural changes in the adult visual system—questions that implied visual potential— were met with quick glances around the table and unsatisfying answers. But I found plenty of unanalyzed data between the lines and in the off-the-record stories told by the neurophysiologists, the men in lab coats. They admitted that, yes, structural changes in the visual cortex happened under certain conditions in animals well beyond the critical developmental stage when, according to the canon, major changes were no longer possible.

I concluded that the neurophysiologists were observing the brain mechanisms underlying the changes I had experienced while doing the Bates Method of vision improvement and while learning to truly see the savanna. I concluded that they were observing the capacity of the visual system to change under certain conditions, such as when one's attention is lit up, and that such change was simply less frequently observed in adult animals than in the young, excited ones, the ones that were still naturally attentive to the world. This didn't mean, however, that structural and functional changes couldn't happen in adults. By this time, I had learned that this kind of oversight is one of the classic limitations of Western scientific methodology. Like lies of omission, science names truth without reminding us that there is more to be revealed. And besides, our scientific methods are simply more interested in norms, or the most common results, than in potential.

So I took visual system change to be fact. I had experienced it, I had watched others experience it while teaching vision improvement, and now I had a neurophysiological picture of it. But because visual system change in adult animals was not published as a conclusion, as a finding, I became suspicious of statistical methods that did not include data representing potential. I presumed that the data indicating potential represented the few adults who were naturally more attentive. I wondered about the methods for "massaging data," notably the only statistical methods we were taught. I began asking bigger questions, questions that focused on the history and assumptions—the entire enterprise—of Western science. I wanted to know why our ability to change the way we see was not big news. Why, I wondered, did our research tradition focus on norms at the expense of identifying the great potential inherent in having sensory systems, in having an exquisitely tuned human body?

I pondered these questions while crying and compromising in order to collect data, pass exams, and write the thesis that would earn me a doctorate in visual science. After two years of data collection, of being buried in numbers to such an extent that entire seasons passed without my notice, I retreated to a quiet coastal cabin to write my thesis and recover my sensibility and sanity. I kayaked at high tide each day, for three or four months, and remembered my alignment with the tides, with the great turning of the earth. I took off my glasses and began to see the world again. I saw egrets, osprey, and herons; the exact moment the tide reached its highest point; and the sunset—crimson, brilliant, glorious, and different every day. I saw stars so vast I wanted to throw myself into them. And because the mornings were so profoundly still, precious, and serene, I saw myself be nudged beyond sanity into joy and great pleasure.

Several years later, while kayaking in Desolation Sound, I noticed a nasty blind spot in a friend. I had always considered her to be perceptive and so the incident took me by surprise. Over the course of six or seven days, we paddled past many small islands, many rocky beaches, a few sandy beaches. We saw otters, osprey, and great expanses of bright blue water, often still and reflecting shorelines of dense green fir. We found freshwater creeks slipping into silent coves or into little bays where there had to be secrets and stories. We swam many times each day, diving off rocks and dropping ourselves into warm, salty water. In each moment, our place was marked vividly by sunrises and sunsets, by hidden beaches, distant mountain ranges, and sharp, white peaks piercing brilliant skies—by the infinite distinctions of land and sea. We paddled in a wide arc, circling the southern portion of the sound. As we passed places seen and touched days earlier, my friend asked, "Have we been here before?"

My mouth must have dropped open, so stunned was I by her question. Like many of us, she had not seen where she was. She had not noticed the relationships that so fully marked our place.

This book is most essentially about seeing where we are. The relationships between things—between the peak of a mountain, the place of sunset, and the silent cove to the south—show us where we are. For navigators, this is locating one's self by means of triangulation, a relational perspective. The relationships between ourselves and Mountain, or the way we do or do not resonate with the sunset, show us "where we're at" on the internal landscape or in psychological terms. But because we have turned our focus of attention toward material things, toward static objects and unrelated possessions, the world of relations has slipped from view. As a consequence, we do not truly see either where we literally are or the truth of "where we're at." We do not fully see the relationship between the disappearance of frogs, on one hand, and the paving of wetlands, on the other hand. Nor do we readily see the large-scale relationships between planetary health and personal experience.[3] For example, with our eyes closed to the loss of frogs, we also miss the color and vibrancy of the world and the pleasure that naturally arises from a fully sensing human body. We miss sensing our engagement with sounds and scents, the sensuality of our bodies, and the feeling of belonging within a vital field of

relations. It becomes a personal loss, "the extinction of experience."[4] And so we suffer from a form of loneliness—as if we no
longer know our home, or who the relatives are.

This is a book about reclaiming our vision in the largest sense.
It is about cultivating our sensitivity, the senses learning to gather
up the scent of soil and the subtle touch of tongue, to see dense
color and rich texture, to experience the saturation and depth of
the world. This book is about awakening and recovering the fullness of our sensory, sensual, and perceptual capacities, about the
very lusciousness of having senses. It is about fully perceiving the
Others, whomever they may be—tree, lover, or the cat next door.
This book is about revealing the nature of our relationships with
them and cultivating a wild love for the local characters—the jays
jabbering outside the window, the ponderosa pines nearby, the
narcissus just blooming. Through the senses, they get under the
skin, gathering in the body and spilling into the heart.

It is *only through the senses* that we experience what it means to be
fully human. It is only through the engaged senses that we are able
to feel desire and intimacy, the great longing to be fully, wholly,
and utterly *in* the world. But it is only through the *honest and engaged*
senses that we will come to appreciate the living world as it truly
is, both wildly beautiful and endangered. Cultivating our perceptual capacity is fundamentally related to both the quality of our
personal lives and restoring the quality of life on the planet.

There are perhaps a bazillion stories about different abilities
and ways of seeing, stories that demonstrate the tremendous potential within the visual system. For example, we have the capacity to see the light of a candle at a distance of several miles. We
may learn to discriminate between wavelengths, perceive mi-
croorganisms without a microscope, and see shadows in the light

of a new moon. Tom Brown, a well-known tracker, sees tracks on hard rock surfaces. By looking at tracks, he knows if an animal is hungry, if she has just eaten, if she is newly pregnant. I am told that he can tell from a single set of tracks if there has been a miscarriage, if she is troubled or uncertain, if she is about to run or change direction. At age fifty-something, my friend Walt, a naturalist, says he can see better now than ever before. After doing his annual day-long bird count in the Bradshaw Mountains he called me to say, "Life!, life!, so much life! How can we not *love* the world?!" No doubt, he was literally bursting with the life he had seen and absorbed all day. In the realm of subtle sensation, the Kalahari Bushmen know when a visitor is arriving, still several miles away, by feeling a vibrational change in their hearts. In the realm of knowing one's place, Charles, the Tanzanian with whom I did research, would make a beeline back to the jeep at the end of each day, after eight hours of following baboons who constantly zigzagged, circled, and changed direction. He never used a compass and we were never lost. It was a matter of seeing where he was. Charles could also, from half a mile across the savanna, name any of the 120 baboons we followed. He could predict, from a subtle shift in the color or texture of a baboon's skin, the exact day on which she would be most fertile. And like Tom Brown, he could, in fact, see every phase of the female cycle. The stories about Charles's perceptual prowess are many. The stories about the power to perceive are endless.

Vision is a power. In shamanic traditions, vision is commonly believed to be the power to "catch" or receive images and visions and the power to "weave" together the human imagination with the sensible world. These are interdependent psychological processes, with the ultimate power of vision being the ability to co-create the world we see and act upon. Our actions upon the world further our co-creative influence. We reach out with our hands, with our choice, to touch and transform what we see. The world is thus influenced, handled and shifted a bit. This makes vision a true power—particularly when it is a shared vision.

In contrast to the earth-based, shamanic traditions, Western psychology defines perception as "becoming aware of by means of the senses" or "the process by which sensory information is organized and interpreted by the brain." Although these definitions are accurate, they say nothing about creative power and little about having influence. These definitions highlight the "organizing and interpreting" that we do, implicitly referring to the fact that we carry some responsibility in the way we see the world, but this responsibility is not readily apparent in the definitions. In other words, Western science defines perception in relatively passive terms. The definitions say nothing about our intentional or embodied involvement in the process of seeing.

As an alternative to the scientific view, perception is increasingly thought to be an active engagement between ourselves and the sensible world. David Abram, an environmental philosopher and the author of *The Spell of the Sensuous*, says, "Perception is not a cerebral event but a direct and reciprocal exchange between the organism and its world."[5] Yi-Fu Tuan, a geographer and the author of *Topophilia*, defines perception as "a reaching out to the world." According to Tuan, our capacity to perceive is minimized by passivity. Rather, he says, "the fingers must slide across a surface to feel," and "the eyes must roam across a landscape to see." He continues, "It is possible to have eyes and not see, ears and not hear."[6] In other words, perception is a participatory act. Ralph Abraham, a complexity theorist, defines perception as the way in which human consciousness interacts with the electromagnetic field, or light. With reference to the co-creative power of vision, he adds that the easiest thing to affect by the power of mind, or one's intention, is the electromagnetic field.[7]

This book is about the nature of our participation with the visible world. Are we participants or passive receivers? Does a sense of co-creative potential accompany our open eyes? Do we perceive the world with an eye toward acquiring more possessions? Is the world full of dead stuff for our consumption or is it alive and animated? Do we imagine a world soul with whom we might interact? And with reference to the shamanic traditions, how do we weave together our own psychological powers with those of the natural world? How do we both deepen our individual experience and co-create a healthy world for our children?

Modern consciousness, for reasons I will elaborate on, is conditioned to think in "either/or" terms. We tend to think of reality in black-and-white scripts. Given such habits of mind, we unintentionally separate ourselves from the world, forgetting that we are in the midst of a flood of penetrating light, that we are saturated by the sounds and scents that tell us who and where we truly are. We forget that these sensations directly influence our psyches, our bodies, and our very identities. In every open-eyed moment, we are threaded through with sensation, shifted by a sign, or contracted by a sound. We cannot help but react and respond. We cannot help but mix and merge our habits and hopes with the signs and signals of wherever we may be. But what are the habits and hopes? What is the quality of our presence as a wild wind wraps around us, or as a lover's lips touch our own?

To me, as a kind of mystic scientist, perception is the dynamic ground of our many relationships with the world. Perception is the energetic movement between what is inside of us and what is out there in the world. From the inside, perception is the act of receiving, transducing, and transforming vast amounts of energy. Visual perception is the transduction of electromagnetic energy, or light, as billions of photons and vibrating waves move into us. The light is transduced and translated into what we see, believe, and act upon. Depending on the quality of our presence and the receptivity of our bodies, perception may vary between extremes. We may, in some fundamental sense, be "blind"; our relationship with the world is thus minimized. On the other hand, we may be "perceptive" or profoundly sensitive in the moment of opening our eyes. And if the incoming streams of light, sound, and scent coalesce with a presence of mind and a truly participating body, perception may become the ground for a sensuous, even ecstatic, relationship with the world.

We often say, "Seeing is believing." From a psychological point of view, it is equally obvious that we tend to see what we believe, to project our private worldview upon the surfaces of the world. Do we believe in the possibility of an ecstatic relationship with the sensible world? Do we imagine *loving* the world we see? Our beliefs weave through our sensations and tend to appropriate reality in the interest of our own version of how the world is or should be or, perhaps, *could be.* Because seeing determines action, the recognition of both our projections and our internal powers to co-create the world has significant implications for the quality of our lives and for the quality of life on the planet. James Hillman, a radical Jungian psychologist, simply says: "By seeing differently, we do differently."[8] And by believing differently, I must add, we see differently.

In a practical sense, this is a perceptual guidebook for deepening our experience. Deepened experience implies a relational way of seeing and a relational way of being. A student of mine says that deepened experience is "that quality of *sensuous and honest* engagement in which we expand to include all the living legends," all the possibilities inherent in having a life. From a *sensuous* perspective, deepened experience is full of desire to be *in* and *of* the world, the earth, the sounds and scents that stream into our bodies. From an *honest* point of view, deepened experience is about looking at all of it, the whole world—the beauty and the beast.

In addition to being a guidebook, this book is a prayer for the planet. My prayer is that we learn to see, hear, taste, touch, and feel with great desire and care. It is my hope that a return to our senses, a renewed depth of experience, will inspire us to bring our perceptual powers to the co-creation of a vibrant world. I pray for a revolution by force, by the great forces of the universe entering and shifting us, passing between and into our cells, fingering our sinew and shaping our souls. I pray that the untamed world reaches deeply into us—whispering, storming, and breaking open our tender, crusty hearts.

Chapter 1

Varieties of Visual Experience

The science books explained the causes and effects. But old Grandma always used to say, "Back in time immemorial, things were different, the animals could talk to human beings and many magical things still happened." . . .

Dragonflies came and hovered over the pool. They were all colors of blue—powdery sky blue, dark night blue, shimmering with almost black iridescent light, and mountain blue. There were stories about the dragonflies too. He turned. Everywhere he looked, he saw a world made of stories, the long ago, time immemorial stories, as old Grandma called them. It was a world alive, always changing and moving; and if you knew where to look, you could see it; sometimes almost imperceptible, like the motion of the stars across the sky.

—Leslie Marmon Silko, Ceremony

Multiple Eyes

Reality is not what it is.
It consists of the many realities which it can be made into.

          —Wallace Stevens

To see" is a relative term. Its meaning depends on one's cultural context, all sorts of personal factors, and, of course, on whatever one is looking at. Almost any slice of the world may be glittered or dull, rendered by the eye as diamond-like and vibrant or dead—a done deal.

Perception is a radically subjective process, which is to say that we have a great deal of influence on how it turns out. In any moment and place, there are many ways of seeing the world. More precisely, there are many perceptual worlds filtering through the varied lenses of culture, gender, and personal experience. No two ways of seeing are just alike, and yet there are shared views within cultures and within various epochs. These views are firmly glued together by consensus and reinforced by the assumptions, expectations, and traditions of each culture's social system.

On a personal level, we strive to affirm our view of reality, whether it be a momentary perception or a long-term assumption about the way of the world. In psychology this tendency toward affirmation is referred to as confirmation bias. It is a strong effect, showing up as the numerous ways in which we do not fully apprehend the multidimensional nature of reality or, for that matter, the relativity of perception. Instead, we look for and attend to the information that will support our beliefs, as if gathering up our very own view of the world. Such a view is safe. It is a done deal.

It is fair to say that most of us have a limited yet comfortable view of the world. Things fall into the categories we have learned and our lives unfold in primary colors and monotones. But this need not be the case. There are many eyes, many viewpoints, through which to see the world. Saint Bonaventure, a medieval mystic-philosopher, described three such "eyes" with which the world may be perceived. According to Bonaventure, the "eye of flesh" perceives the material world, the world of objects; the "eye of reason" apprehends the mental realm; and "the eye of contemplation" reveals the transcendent realities. From Bonaventure's point of view, all knowledge issued forth from these three "forms of illumination." We might add one further view to this. James Hillman, author of *Re-visioning Psychology*, suggests that soul is that which deepens events into experiences.[1] If stories are events coalesced into experience, then Leslie Marmon Silko's character in *Ceremony* is seeing dragonflies with soul, with the "eye of soul."

In fifth century B.C.E. Greece, there was a good deal of elevated discussion about the nature of reality. At the time, two different views of reality were being debated. The old school, which we might call naturalism, was faithful to the realm of the senses, to the shifting, ever-changing nature of the visible world. But with the birth of philosophy, a kind of rationalism was rapidly emerging as the new school of thought. The debate was really between the pluralism apparent in the sensory world and the absolute, unchanging nature of reality—the absolute Truth. Naturalism implied never knowing for sure, giving one's trust to the uncertain realm of the senses. Rationalism implied the divination of order, of a transcendent, unifying reality that alone held the fundamental truth. Empedocles, a philosopher of the time, heroically attempted a reconciliation of the two disparate views. He suggested that earth, air, water, and fire were unchanging and sacred elements that, although invariant, constantly shifted in their relationship to one another as they moved together and apart—influenced by the primary forces of love and strife. The relationships between the elements changed, but the physical matter itself stayed the same. Empedocles clearly chose both naturalism and rationalism, as if stepping around the temptation to have a single definitive answer regarding the nature of reality.

Empedocles' philosophy included an early theory of vision. He postulated that Aphrodite, the Greek goddess of love, created the eye, the experience of seeing, by "riveting" the four sacred elements together with love. Aphrodite then "kindled the fire of the eye at the primal hearth fire of the universe" and formed passages for transmitting the eye's "fine interior fire" to the world. Thus, our seeing was born of love, conceived of love. Such was sight. According to Empedocles, "Sight proceeded from the eye to the object seen; the eyes rayed out their own light."[2] Perhaps most notable in Empedocles' theory of vision is the idea that the proper *correspondence* between the intraocular fire and the external fire was essential for vision to occur. Sight required resonance, a vibratory relationship between the seer and the seen.

When I imagine Empedocles debating with his contemporaries, I envision him working carefully, like any serious philosopher, to develop his argument. I imagine that a thoughtful kind of faith wove together his notions of love and the eye. I look back across centuries and wonder at the source of his faith; *what was he seeing?* Sight born of Aphrodite, I can only imagine, must be akin to looking at the world through "love eyes," both intently focused and soft, both present and profoundly receptive. Through "love eyes," magic happens, events coalesce into synchrony, and like the lovers who forget to eat, even food recedes to the background— our eyes feasting on the beloved.

Empedocles' vision must have glittered or glowed. But it was Plato (427–347 B.C.E.), born less than a hundred years after Empedocles, who actually set the stage for many of our habitual Western ways of seeing. Plato was also a kind of mystic, although his theory of perception edged us much closer to a rationalist worldview. For Plato, vision occurred when a stream of fire emanated from the eye and coalesced with daylight, forming a unified and "harmonious body of light." Plato believed that the emanating "fire of the eye" was "a light of interpretation, of 'intentionality,' . . . a light that grants meaning."[3] For Plato, acuity meant seeing the *ideal* image. The ideal was an archetype, endowed with a kind of divine and universal significance. To see a horse ideally meant to see "horsiness"—not the particular muscle and mane. "Horsiness" was to be found in the clarified mind, not in the horse itself. The archetypes borne of the mind's transcendent intelligence constituted absolute truth; the sensible world simply pointed toward what Plato called the "Ideals." With this inward turn of focus, Western sensibility began to see the human mind as the ultimate source of the absolute and ideal, and the sensible world, the world outside of ourselves, began to recede into the background.

The Stoics of the second century based their theory of vision on the conception of pneuma, an ever-present ether composed of both air and fire. The optical pneuma—the ether that arose in the center of consciousness and flowed through the eye—"stressed" the air at the interface between the eye and the world outside ourselves. Vision occurred when the stressed air, touched by the optical flow of pneuma, was illuminated by sunlight. The Stoics believed that the activity between the optical pneuma and sunlight represented a transformation. In the moment of seeing, pneuma and light, now twined together, became an instrument of the soul. Vision, then, was both an extension of ourselves and a soulful engagement. The power of vision resided in its ability to co-create reality.

Galen, a second-century physician, summarized his era's two principal theories of vision in a single statement: "A body that is seen does one of two things: either it sends something from itself to us and thereby gives an indication of its peculiar character, or if it does not itself send something, it waits for some sensory power to come to it from us."[4] In other words, the energetic flux essential to vision originates and emanates from either the viewer or the viewed. In the ninth century, Al-Kindi, an Islamic philosopher, further summarized this idea by suggesting that, like the eye, everything in the universe emanated light; the world was, in essence, a vast woven web of light-like rays.

The notion that *we* were the source of this energy—that the perceiving eye illuminated the world, that an ocular fire emanated from the seer—was shared by numerous Greek and Islamic philosophers and persisted for some fifteen hundred years. In the West, this theory was taught into the twelfth century. The central theme of this school of vision was the fundamentally active and formative nature of the viewer's role in perception. During the middle thirteenth century, Saint Bonaventure's three ways of seeing implied the power of intention. Depending on which eye one brought to any perceptual encounter, the world was composed of objects and appearances, mental images, or transcendent, ultimate realities. But Bonaventure added a crucial component to his theory of vision. As minister of the Franciscan order, he wrote one of the great spiritual books of all time, *The Itinerarium*, or *The Journey of the Mind to God*. The journey consisted of six steps, or "powers of the soul," that led one toward union with the Divine. The first power was explicitly concerned with the proper use of the senses:

> For creatures are shadows, echoes, and pictures of that first, most powerful, most wise, and most perfect Principle, of that eternal Source, Light, Fullness, of that first efficient, exemplary and ordering Art. They are the vestiges, images, and displays presented to us for the *contuition* of God, and the divinely given signs wherein we can see God. These creatures are exemplars, or rather illustrations offered to souls as yet untrained, and immersed in the senses, so that through the sensible things that they see they may be transported to the intelligible which they do not see, as through signs to that which is signified.[5]

The things of the world were signs and images of God, to be seen with the engaged powers of perception.

The Sufis, like Bonaventure and early Greek and Islamic philosophers, also recognized perception as a form of creative power. Given this power, the world is colored and textured with metaphor and divinity. Through "the eye of the heart," Sufi masters see the world as if "cloaked in celestial light."[6] The world becomes "God-in-disguise, God veiled." Objects are "garments that God dons to create a world.[7] According to Sufi teachings, as the eye of the heart grows stronger the garments become increasingly transparent, increasingly revelatory of God.

On the other side of the planet, Australian Aborigines see through "strong eye," or through what I assume to be the power of emanation. According to tradition, the power of strong eye is drawn up from the center of the body, just below the navel, where a rainbow serpent is said to be coiled and resting. With practice, the power is drawn into the third eye in the center of the forehead. The force of strong eye is then rayed out to the world. From this perspective, Aborigines claim that they are able to discriminate between truth and dishonesty, diagnose illness, penetrate another's mind, and transfer thoughts.[8] Strong eye is practiced, assertive, and sees *into*.

For Australian Aboriginal peoples, acute perception was once a given. As hunter-gatherers, visual acuity—or, more precisely, acumen—came with the territory, for one's subsistence depended on the power of the senses. Traditionally, daily life included wandering and searching, catching a flash of feather or fur, sniffing, touching, and listening with an intensity that determined dinner. According to Robert Lawlor's remarkable account of traditional Aboriginal life, *Voices of the First Day: Awakening in the Aboriginal Dreamtime*, "Their life in the *yuti* (perceivable world) is one of intense observation of the specific detail of every aspect of nature. . . . They intensely observe any bird, animal or insect in their vicinity,"[9] breaking stems, tasting and smelling sap, sometimes rubbing it on their bodies to discover if it is edible. They also see Dreamtime beings; the Aboriginal landscape is threaded through with the presence of the ancestors and their stories. Lawlor writes:

Every land form and creature, by its very shape and behavior, implied a hidden meaning; the form of a thing was itself an imprint of the metaphysical or ancestral consciousness that created it, as well as the universal energies that brought about its material manifestation.[10]

The seamlessness between interior and exterior landscapes, between the material and symbolic realms, is central to what white explorers named "the Dreamtime." Environmental philosopher David Abram writes that the Dreamtime is "a kind of time out of time, a time hidden beyond or even *within* the evident, manifest presence of the land."[11] It is the continual *becoming* experienced when one is in daily, embodied reciprocity with the land. In such a fluid state of being, the delineations between the apparent and the yet-to-be, the visible and invisible, become irrelevant. Having a proper, reciprocal relationship with the visible and invisible powers means conversing with springs, boulders, and birds, singing to navigate along ancestral tracks, following the songlines laid down by the creative ancestors, and honoring the specific intelligence of each place, each camp along one's travels. Given such hints of the Dreamtime, of a worldview so radically different from my own, I can only speculate on Aboriginal ways of seeing. I do know, however, that young Aboriginal children run circles around the best Western trackers.

In *The Primal Mind: Vision and Reality in Indian America*, Jamake Highwater describes the American indigenous perceptual mode as "sympathetic," a form of perception that fully recognizes the presence and value of "the thing seen." The object, person, or place is seen as one of a "multi-verse of possibilities." The thing seen displays a quality or way of being—the stillness of a mountain, the rootedness of a ponderosa pine—and suggests the possibility of "temporarily turning into" something other than one's self. What Highwater is suggesting is the capacity to *become*, to try on the attributes of another, of whatever is displayed or modeled for us in the visible world. Seeing the world with an eye toward *becoming* teaches us about the very nature of transformation. In any culture, transformation—whether viewed as a rite of passage, a developmental stage, or a form of shape-shifting—is a fundamental psychological process. To have the process displayed in the forms and faces that constitute the landscape is to place one's development, one's shifting psyche, in the landscape. The landscape thus mirrors the state of one's deepest self. The reflections are dynamic, changing with internal shifts, as if landscape and psyche are engaged in an ongoing conversation. This form of psychology is not an analytic process; *temporarily turning into* arises from direct experience, from looking and listening.[12] It is a way of coming to know one's self as firmly rooted in the perception of place, where one actually is. In native psychology, "Who am I?" thus becomes "Where am I?"

According to Highwater, "turning into" requires both a consciousness in which multiple realities have equal value and a particular, generous form of projection. This form of projection is described as "divining the inner life of everything,"[13] or granting a unique individuality to the "thing seen." Again, within the confines of a Western sensibility, we can only guess at what this implies; for starters, we do not dare risk divining the inner life of the Other for fear that we may challenge the canon of objectivity. Second, traditional Western psychology views the psyche as residing solely in the human head. To see our inner world reflected in the landscape is at best a metaphoric way of seeing, is it not? But, I won-

der, what is *this way of seeing?* What kind of consciousness follows
on the footsteps of such sight?

The Shipibo-Conibo Indian shamans of eastern Peru experience a form of hallucinogenic seeing that further challenges our
Western sensibility. In trance, the shamans perceive complex patterns of woven lines. The patterns appear to pulse and float downward through space. When the pattern reaches the shaman's lips,
he sings them into songs for his patients. The pattern is then trans-
mitted to the patient in the form of sound. When the songs reach
the patient, they are transformed back into a visual pattern, which
then penetrates the patient's body and heals. However, it is neither
the songs nor the image but the *fragrance* of the pattern-songs that
ultimately carries the power to heal.[14]

This perceptual possibility is way beyond my experience and
nearly beyond my imagination. My modern, Western mind has little if any reference for such realms of experience. Nonetheless, I
ache for, as Kierkegaard said, "the eye which, ever young and ardent, sees the possible."

Three other ways of seeing are described within the teachings of
Mahayana Buddhism. The first, referred to as *parakalpita*, is the
most elementary. It is a categorical way of seeing that requires little attention or contemplation and is considered to be "false discrimination." From this perspective, "the thing seen" is not truly
seen but rather named and judged for its surface appearance. One
might say that "a tree is a tree is a tree," or "once a tree always a
tree."

In contrast, *paratantra* is a relational way of seeing. For example, the sun and its rays are seen as inseparable—one cannot be perceived as independent of the other. Through this lens, the interdependence of the cosmos becomes apparent. Buddhist monk Thich Nhat Hanh illustrates this way of seeing with a piece of paper: "There is a cloud floating in this sheet of paper. Without a cloud, there will be no rain; without rain, the trees cannot grow; and without trees we cannot make paper. . . . If we look into this sheet of paper even more deeply, we can see the sunshine in it."[15] In *The Sun My Heart*, he says, "Sunshine is green leaves. Green leaves are sunshine. . . . Do you think you can separate sunshine from the green color of the leaves?"[16] The third way of seeing, *parinishpanna*, is considered to be true enlightened perception. This mode of perception sees the (reflected) moon as both in and on the water, neither immersed nor outside it. It is seeing that the moon is both *not in the water*, for it is mere reflection, and *there*, for the moon is truly before us. This is the moon and water unified, a picture of ultimate, non-dual reality. Perceiving ultimate reality is observing the image freshly, independent of what we think we know about it. It is seeing *into*, unconditioned by categorical perception.[17]

Perhaps parinishpanna is akin to *darśan*, a Hindu concept that constellates a sacred way of seeing. Darśan means "seeing" but is also sometimes translated as "auspicious sight," and it specifically refers to both seeing and being seen by the divine presence. In *Darśan: Seeing the Divine Image in India*, Diana Eck says, "Beholding the divine is an act of worship, and through the eyes one gains the blessings of the divine."[18] Darśan is revelatory, illuminating the unseen, the absolute, through the act of seeing the tangible physical world—the places of pilgrimage, the temple images, the peaks of the Himalayas, the river Ganga, the gaze of holy persons, one's guru, the saints, the renunciates. Darśan implies being gifted by the thing seen in a moment when the observer is receptive and respectful, and thus able to truly see.

Perhaps this receptive and respectful form of engagement is what Ken Wilber refers to, in *The Eye of the Spirit: An Integral Vision for a World Gone Slightly Mad*, as "simple, ever-present awareness" or witnessing what is "always already." Ultimate reality, then, "is not something seen, but rather the ever present seer."[19] I have no doubt that "the ever present seer" refers to the energetic stance of the viewer and to the relinquishing of our concept of self as separate from "the thing seen." It describes a form of participating consciousness in which references to self, subject or object, become irrelevant. From such a stance, the seer becomes the seen. This is non-dual experience, the unbroken experience.

So what, then, does it mean to see? Is it, heaven forbid, shifting out of ourselves, relinquishing the only, albeit shaky, ground we stand on? Perhaps the most we can know with any certainty is that perception is relative, that what we mean by the verb "to see" varies across cultures and time periods. The variations might depend on where the energetic flux or intention arises within any perceptual exchange. Is the source of this flux the razor-like emanation coming from the stranger on the freeway, impelling you to turn your head and see what all the energy is about? Or perhaps what catches your eye is the emanation that arises from a stone you see on a sandy beach. You bend down, pick it up, look, gaze, maybe wonder why *this one?* Why did *this one* catch your eye, as if beckoning for your attention? We all experience this, the call of a particular purple flower, an autumn leaf, a single stone.

My basic question is, are we on the receiving end of the energetic flux or are we the energetic source of the exchange? Does the source of the emanation determine what is ultimately seen? Could it be that we see differently, depending on whether it is the viewer or the viewed that emanates energy with greater force? Does a particular quality of visual fire, a transmission or projection, constitute a way of seeing?

What we see clearly depends on which "eye" is doing the seeing. The sight of an "eagle eye"—vision that is directed and precise—is certainly distinct from what one sees through "the eye of the heart," which the Sufis describe as a kind of soft and glimmery view. Across the spectrum of what it means to see, the expression "the eye of . . ." implies that seeing is relative to the inner realm, the place of the eye, inside. Shifting our belief about the nature of vision from one that is largely passive to one that includes our own contribution, our own ocular fire as a source of seeing, suggests that the realm of ideas and imagination is a part of perception. "Ideas give us eyes, let us see," says James Hillman in *Re-Visioning Psychology*. "Moreoever, without them we cannot 'see' even what we sense with the eyes in our heads, for our perceptions are shaped according to particular ideas."[20] If we believe in emanations, if our beliefs include the notion of an ocular fire, then the quality of our gaze—the hunger and intensity of it, the desire of it, the spirit and ease of it, the assumptions that accompany it—is woven into the world.

One thing is certainly true: We do not imagine multiple ways of viewing reality—or the potential implied by multiplicity—unless we have considered what it means *to see* in the first place. In other words, our ideas about vision—whether well considered or taken for granted—are intimately entwined with how and what we actually see. Our ideas of "to see" and our actual potential to sense the world are inseparable; they are psychologically interdependent.

Both historical and cross-cultural "sensory profiles," or glimpses into other perceptual realms, suggest that a full definition of "to see" must also include an appreciation for the many ways of seeing. Given a vast interiority (the endless realm of ideas and imagination), and a still vaster exteriority (the realm of all that is outside of ourselves), we have the potential to co-create a tremendous range of realities.

The great variety of sensory experiences are all different forms of translation between inner and outer worlds. Perception represents the meeting place of interiority and exteriority. Of course there has been considerable debate about where this meeting happens. For the long tradition of vision theorists stretching between Al-Kindi, the great philosopher of the Islamic world, and Galen of the second century, perception happened outside ourselves, as the emanation of the ocular fire met sunlight or the rays of the "thing seen." In modern science, vision happens as light penetrates the eye and meets up with other neural activity in various parts of the brain. Because there are many fibers sending signals down from the visual cortex to the lateral geniculate nucleus (LGN) in the center of the brain, and because the LGN is a "relay station" in which many neurons exchange signals, Francis Crick theorized that the meeting place between exterior and interior worlds happens in that part of the brain. Those who believe in the adage that beauty is in the eye of the beholder would seem to agree with such an unlikely pairing of both the passive reception of signals and high degrees of subjective influence.

Our modern Western assumption then seems to be that the act of perception happens entirely in the human head, as if the sensible world is little more than a trigger to begin the perceptual process. But could it also be true that the thing seen emanates something that catches our eye, that beckons and calls us out of ourselves? Could it be that we truly see when we reach back to "the thing seen," emanating our own fire, our very presence, now resonating with recognition? Could it be that beauty beckons to us, causing us to involuntarily gasp—to literally in-spire—and to be inspired to identify, to give ourselves over to the scent and sight of a brilliant rose glistening in a brilliant morning?[21]

It is my belief that in the act of perceiving, our intention and imagination are cast both outward and inward to varying degrees. Consciously or unconsciously, we choose where the marriage between the inner and the outer worlds occurs.

Two Eyes

> Objective thought is unaware of the subject of perception.
> —*Maurice Merleau-Ponty*

Some would say that there are essentially two ways of seeing, subjectively and objectively. The first is heavily flavored by personal habit and personal pleasure and is dangerously close to living within the realm of illusion, under a smoky veil. At the other end of the spectrum, the objective way of seeing is ideally free from the transient influences of need and desire or the coloration imposed by preconceived notions. It is coolly distant and disengaged from the observed object.

Objective and subjective ways of seeing represent the two poles of the conventional Western worldview. An objective stance is one in which the observer steps aside according to a set of rules. We take ourselves out of the picture, observing reality under carefully controlled conditions in order to see exactly what x quantity causes what y quantity, and with the hope that we might entirely avoid the hazards of subjectivity. We know from quantum mechanics, however, that objectivity is never entirely possible. The "observer effect" refers to the fact that, at the quantum level, an experimenter's observation influences the result of any experiment. Nonetheless, the modern mind continues to place a great deal of faith in objectivity, particularly when reality is viewed through the lens of science.

Subjective sight is the other side of the coin. It is felt, ardent, arousing, and the reality of the perceptual process. Trust in our subjective experience tends to be diminished by our history and cultural codes of behavior. Nonetheless, subjectivity blazes across the screen of awareness when we least expect it—in romantic, rhapsodic moments, when a scent suddenly transports us to a particular spring day, to a remembered love. Subjectivity is visceral, threaded through with story, woven with wind, embellished with desire. Visions freely infused with subjectivity are deeply informed by our expectations and imagination and are especially colorful. But when seduced by the potent power of subjectivity, we have to ask ourselves, what's real? What can we trust as consistent across this fluky realm of subjective sensation? Can I trust the friendly tone and good looks of the man I just met? Is it his sincere desire to know me or my own history and hopes that embellish his voice? As always, this way of seeing slips from certainty.

Science claims to provide us with the certainty we long for. According to Morris Berman, a cultural historian, modern science emerged in the seventeenth century with the question: "How?" The seer became the detached observer looking for a prescriptive answer, and the process of "to see" emerged as more mechanical than soulful. In contrast, the medieval mind (just prior to the emergence of seventeenth-century rationalism) was continually engaged in a search for meaning. The predominating medieval question was "Why?" and the predominant intention was the satiation of one's soul. The two poles of subjectivity and objectivity—here represented by the questions "Why?" and "How?"—appear, disappear, and reappear in the history of Western civilization. Although the objectivist, rationalist approach has long dominated our perceptual habits, subjectivity has had its determined proponents and eras throughout the history of Western thought. The medieval era, in particular, represents a historical period in which subjective sight was given full, albeit unconscious, credence.

The medieval worldview was rooted in symbolism. According to Berman, "Things were never just what they were, but always embodied a nonmaterial principle that was seen as the essence of their reality."[22] Furthermore, the medieval mind understood that the entire universe—excepting God—was continually in the process of becoming. The guidance for "becoming" was to be deduced from direct observation of the organic, animated universe, permeated with signs and symbols. The Doctrine of Signatures, considered the foundation for much medieval thought, banked on the currency of resemblances. For example, because the wrinkled surface and shape of walnuts resemble the human brain in appearance, they were to be ingested for head ailments. The doctrine presumed that the function of any life form was apparent in the visible form itself, that purpose and significance were visible to the observing eye. These "signatures" were ubiquitous, everywhere informing the human psyche. For this reason the medieval mind was involved in searching out the world, continually observing signs that might deliver the viewer to an elevated existence. Perception thus arose within the context of "participating consciousness," a fluid immersion in the organic unfolding of the world—so much so that self and Other lost distinction. As Berman writes in *The Reenchantment of the World,*

> Participation is self and not-self identified at the moment of experience. The pre-Homeric Greek, the medieval Englishman . . . and the present-day African tribesman know a thing precisely in the act of identification, and this identification is as much sensual as it is intellectual. It is a totality of experience: the "sensuous intellect."[23]

Like Bonaventure's first step in "the journey of the mind to God," the Doctrine of Signatures assumed a divinely coded relationship between the seer and the seen. As an active element of consciousness, it fostered an animistic way of seeing, for the world was alive with meaning and message; one participated or "conversed" in order to reveal the meaning inherent in the visible form.

In stark contrast to the sensual participation of the medieval mind, the radical revolution in consciousness of the seventeenth century—marked by the Renaissance and the scientific revolution—gave way to the celebration of objective, nonparticipatory perception. Such thinking produced the camera obscura, the precursor of the modern photographic camera. The camera obscura, truly a black box analogy to the eye, implied that vision consisted of passively registering incoming signals to be processed mechanically—devoid of an inner eye, ocular fire, or even a curious participation with the world. It was Leonardo da Vinci who suggested that the eye was no more than a dark chamber for receiving the signals of the world, and it was Johannes Kepler who developed a geometric and finely framed optical explanation for the camera obscura.

Both da Vinci and Kepler, needless to say, had influence, and the camera obscura became the dominant model of the visual process. By definition, this model insisted on an objective truth. Cast within the growing zeitgeist of objectivity and scientific fact, the inherent subjectivity in the visual process became irrelevant.

Against the current of eighteenth-century science, Johann Wolfgang von Goethe sought truth through a form of practiced perception. For Goethe, truth was revealed by the attentive engagement of the viewer, by the interpenetration of perceptual and cognitive realms. The thing being seen was to be observed, to be looked at again and again, and the observing mind was to be in a state of open, respectful inquiry, searching out the meaning. Goethe's sensory participation transcended the eye in his effort to "see into" and to develop, as he called it, "an exact sensory imagination."[24] The imagination was cast outward, in honor of the actual image before one's eyes. It was to be true to the Other. His way of seeing was both intuitive and rigorous, an intentional form of engagement in which the body, as the sensory system, was receptive and engaged, searching for patterns, for the order and hidden meaning, the "exalted perceptual experiences."[25] As Theodore Roszak wrote:

> He worked from the qualities—always from the qualities: color, texture, above all form . . . the sweet nourishment of the senses. . . . His eye for form and color was almost voluptuous; it caressed what it studied and felt its way in deep.[26]

Goethe's process of *seeing into* began by "plunging into seeing," by removing one's attention from the verbal, intellectual realm and giving it entirely to the thing seen. The internal focus of attention was to shift from the analytical mind to the imaginal realm, embracing both the image within one's mind and the one before one's eyes. It required observing the phenomenon all at once, in its entirety, and finally reaching a kind of active crescendo in inquiry, a questing of the observed phenomenon: What is that color, that brilliant, deep red? What *is* that color? What is that *red color* of *this* rose, the petals spiraling into the center, down, like the color descending deeper into the body. Of what does the rose speak? Of what sensuous fare? Of what higher ideal? Of true, deep, and impassioned love? In looking into a rose, I remember—through my senses—that the red rose is the classic expression of true love. In some sense, in the coalescing of the senses, the rose displays the very ideal of love.

In Goethe's process of seeking the perception of ideas, new "organs of perception" were developed, capable of beholding, as Arthur Zajonc says, "the ideal within the real as archetypal phenomenon."[27] Was this referring to subjective or objective sight? Could it be beyond the dichotomy of the question? Could it be what Abraham Maslow termed "fusion knowledge" or a "caring objectivity"? This way of perceiving is reminiscent of the Doctrine of Signatures, with the added intellectual rigor that was rapidly developing in the modern mind. We might also imagine Goethe's method as a variation on the combined themes of Aristotle, in which archetypal, true knowledge was to be derived from the experience of the senses, the original form of empiricism, and Plato, whose apperception of the Ideals through careful mental development was the original rationalism. Goethe's engaged practice and theory of perception represents both the synthesis and the repetition and recycling of perceptual theory, of the odd struggle to know the truly uncertain nature of our relationship with the ever-changing sensible world.

But Goethe was essentially alone in his ideas of perception, and the popular objectivist view proceeded in its contrary development. Although vision was emancipated from the black box of seventeenth-century rationalism by the nineteenth-century scientific community, this liberation came with a new twist. The modernist perspective restored a form of subjectivity, but solely in the interest of the human body. With much enthusiasm for the burgeoning science of physiology, the viewer's production of vision in the body—the physiology of vision—became all-important, and the "thing seen" receded to the background. Although vision was once again located in the subject, the process remained mechanical and the object of perception became no more than one of the many inanimate things that began to fill a rapidly emerging materialistic culture. Ironically, the developing theory of perception, enamored as it was with the mechanics of physiology, left the body empty of sensation, of a truly embodied experience—what we might recall as sensuality. Thus, through historical maneuver too convoluted and lengthy to fully describe,[28] both viewer and viewed became stripped of "fire" and vision became a separatist enterprise, autonomous and disembodied.[29]

Enter Merleau-Ponty, a French phenomenologist of the mid-twentieth century. According to David Abram, Merleau-Ponty's view of the body as "the very subject of awareness," as the ultimate source of vision, is anything but disembodied: "The sensing body is not a programmed machine but an active and open form, continually improvising its relation to things and to the world."[30] And according to Merleau-Ponty, the body is fully embedded within the "flesh of the world."

From this perspective, perception is a reciprocal process, radically participatory, communal, sensual, a process in which the body is "magically invoked" and "possessed" by the Other. Borne in the body, perception is primary, prior to cognition, and in diametric opposition to the separatism that was emerging from a heady enthusiasm for physiology and other objectivist influences of the time. For both Merleau-Ponty and Goethe, the "bodiliness of sight" was endowed with the infinitude inherent in relationship. Like Goethe, Merleau-Ponty celebrates an engaged, receptive, responsive, ever-shifting, flexible, and ultimately uncertain sense of reality, a phenomenological way of viewing the world, in which the thing seen has its own significance and reality.

Following in this tradition, David Abram, in *The Spell of the Sensuous*, describes perception as a "mutual interaction, an intercourse, 'a coition, so to speak, of my body with things.'"[31] It is "communion and deep communication." Abram further takes the notion of "to see" out of the human mind, out of the visual cortex, out of a mechanistic history, out of an anthropocentric conscription and into the realm of the biosphere, the relational field. Perception is "more an attribute of the biosphere than the possession of any single species within it." It is "an open activity," a "dynamic blend of receptivity and creativity."[32] Like Merleau-Ponty's radically reciprocal mode of perception, this carefully chosen language kindles imagery of Galenic fires burning between the viewer and the viewed, restoring value and vitality to the "thing seen," to the phenomenon itself.

The word "phenomenology" comes from the Greek word *phos*,

which means "to show itself." It refers to that which is manifest, brought into the light of day. The philosophical implications of the root word are best understood if conceptualized in the context of Greek linguistic structure, within which words are expressed in one of several "voices." The first voice is passive, the third is active, and the second voice is both passive and active. We might thus call the second voice an expression of mutuality or reciprocity.

The etymological source of "phenomenon," *phos* or *light,* was expressed originally in the second voice, implying both passivity and activity. This suggests actively bringing forth into the light of day and also receiving light—becoming enlightened by that which fully shows itself. Thus, to be receptive to light is to be enlightened. Such linguistics weds material and spiritual realms, visible and invisible, objective and subjective—making the seer and the seen inseparable, engaged in a reciprocal process.

A view born of phenomenological appreciation, then, is expressive: Phenomena show themselves, informing and enlightening the viewer in the process. The thing seen becomes a material form of intelligence, speaking in languages coded in contours and color, in patterns and lavender scent. Phenomenology not only restores both "subject" and "object" to the conception of what it means *to see,* but also restores to us, the viewers, the power to converse with the "thing seen."

But there is another, more immediate way in which the oppositional poles of subjectivity and objectivity might be reconciled, or recognized as no more than epistemological inventions. We have two eyes, both left and right. The right eye rays out into the world, whereas the left eye is primarily engaged in the act of receiving, absorbing. Look at any photograph. Look carefully at the eyes caught in a sincere moment of seeing. Look at their expression, the flavor of their engagement with the world. Is there some reality in the possibility that we simultaneously receive and reach out into the world? Is the receptive eye glazed, softened, wide-eyed, love-eyed, or shifty with avoidance? Is the right eye raying back sharply, eagle-eyed? Is it defended, hard, or simply attentive and curious? The question is not left or right, objective versus subjective, as we have been led to believe during the many historical cycles of visual theory. Rather, the question we might ask ourselves is, what is the *quality* of our gaze?

The Visual Science Eye

> May God us keep from single vision and Newton's sleep.
>
> —*William Blake*

Science strives for objectivity, but if ever there was a science that suffers from subjective influence, it is visual science. How can our way of seeing possibly avoid interpenetrating not only the method and the interpretation of results but the very questions asked? How can our questions about how we see not be highly influenced by *how* we see? This question is at once a no-brainer and hard to grasp. Nonetheless, visual science has long claimed to toe the line of reason and objectivity, particularly after Wilhelm Wundt, who created the first visual science laboratory in the last decade of the nineteenth century, was chided for his introspectionist methods. Wundt was criticized for developing a methodology that relied on subjects' reported experience, which is to say, on subjectivity.

Our modern Western worldview is largely colored by the

mechanism and reductionism that define modern scientific methodology and knowledge. In spite of a claim to objectivity, the scientific approach comes with a load of associated presumptions. For example, Western thinkers tend to focus on the parts of any system to the exclusion of the system itself. Even in the biological sciences, cells and their components have long been of primary importance. It is only in recent years that organisms, species, and ecological systems have received comparable attention. Western science has traditionally presumed that the properties of complex systems can be described, quantified, and replicated by describing the operation of the independent parts. This is reductionism in its worst form. Although the methods and assumptions of reductionism still comprise most of modern scientific research, disciplines such as contemporary physics, ecological science, and systems theory have revealed the considerable limitations of purely reductionistic thought.

With even a glimpse of the history of an ideology like reductionism, we begin to see the culturally conditioned filter that shapes our imagination and colors our views on perception. Realizing that we are thus conditioned is the first step toward liberating our perception so that we may see around, as opposed to through, this conditioned lens. In *Catching the Light: The Entwined History of Light and Mind*, Arthur Zajonc explores the Western lens by examining three thousand years of our changing conceptions of light. Just as our theories of vision have moved through cycles of subjective and objective descriptions, our notions of light have alternated between light as divine illumination, "the gaze of God," and a mechanistic description. In our era—and particularly within the dominant arena of science—light is defined mechanically as both photons and waves. We view light as the stimulus for vision and measure it in terms of either intensity (as in the number of photons) or frequency (as in wavelength).

Consistent with this mechanical definition of light, visual science has defined "to see" by measuring the capacities of the visual system. The fundamental measurements are referred to as "absolute thresholds," beyond which, according to psychophysical doctrine, humans simply don't detect light. According to this definition, the absolute threshold for vision is a specific measured quantity of light at a particular wavelength. From this perspective, "to see" is a mechanical event, devoid of spirit, metaphor, or sensuality.

The measurement of the absolute threshold is derived from experiments in which subjects are randomly shown flashes of light at various intensities. In classic experiments the threshold intensity was defined as that intensity at which the subject reported that she saw the flash 60 percent of the time that it was presented. The measurement of the threshold, however, depended on the duration of the test light, where it was presented in the visual field, the size of the test spot, and the wavelength of the light. With many variations on the experimental design theme, the boundary between seeing and not-seeing has been carefully defined by visual science.

I have come to realize that there are numerous other factors that influence the "absolute threshold." As an undergraduate, I often volunteered to be a subject in the vision labs. On numerous occasions, while sitting through absolute sensitivity experiments, it became clear to me that my "threshold" varied more than could be explained by the standard experimental parameters. For example, perhaps half of the time, I definitely saw dim lights that, given the method, would be interpreted as "not seen." Even more significant, there were times when my results—my visual sensitivity—

varied considerably, based on whether I'd had a cup of coffee before the experiment or too little sleep the night before. I also knew
from previous experience, particularly while doing vision improvement exercises, that my overall visual sensitivity was notice-
ably greater after meditating.[33] I began to wonder if my intuition
and my perception were the same thing, different only in the degree to which I was receptive to the energy "out there." In other
words, like everything else in the universe (except, perhaps, the
speed of light), I began to realize that the absolute threshold for
perception is relative.

The scientific community arrives at its definition of "to see" by
averaging the responses of a population of "normal" individuals
across numerous days of testing. In other words, the working definition of "to see" is defined by the *method* of inquiry used within
the context of a *particular* way of seeing. There is little, if anything,
in the method that takes into account either our perceptual potential or any mode of seeing that falls outside what happens in a
darkened laboratory setting. In fact, if a subject in such experiments begins to show improved performance by virtue of participating in the experiment, the data are considered irrelevant due to
"practice effects" and are discarded. The subject is sent away. The
major problem with this is that we never know, from a scientific
perspective, what's possible.

Despite my critique of the way in which perceptual potential is,
ironically, made invisible by the methods used in visual science, I
have no doubt that the visual process is essentially the transmission of signals down neural tracks. A bundle of nerves carries signals from the retina to the lateral geniculate nucleus where
"top-down" signals from an individual's cortical storehouse of assumptions, memories, and expectations meet the incoming stream

of signals. After complex mixing and matching of the "bottom-up" and "top-down" signals, the signal proceeds to the visual cortex at the back of the brain, then spreads forward through numerous areas identified as centers for further visual processing.

By the 1980s, the big item in vision research was the investigation of the developmental process in the visual system. David Hubel and Torsten Wiesel, who received the Nobel Prize in 1981 for mapping the "functional architecture" of the visual cortex, set the stage for a flurry of neurophysiological enterprise. They delineated six layers of cortical neurons receiving signals in an orderly matrix, identified ocular dominance columns of neurons, and discovered orientation selectivity in individual neurons.

Most important, their research began to reveal the ways in which visual function and neural substrates are shaped by experience. For example, they found that kittens raised in cylindrical and vertically striped worlds walk blindly into horizontal forms. When tested, the corresponding neurons in these kitten's cortexes are unresponsive to horizontal stripes but will fire enthusiastically when presented with verticals. All of this was exciting news to the vision community, and there were many variations on this research theme. The fundamental finding was an undeniable and direct relationship between visual function, or a way of seeing, and the environment within which cats and monkeys are raised. This is one of those particularly beautiful examples of reciprocity between an organism and its environment. In a functional sense, the reciprocity represents a kind of coevolution in the developmental process.

The relationship between an organism's environment and its visual capacity has profound implications for the way in which human beings learn to see within an industrialized world. Industrialized culture is such a significant creative force that, if we are believers in developmental and evolutionary co-creation, we might simply recognize it as *the* environment. Jerry Mander, author of *In the Absence of the Sacred: The Failure of Technology and the Survival of the Indian Nations*, argues that we "are essentially living in our own minds"[34] by virtue of the human-created and human-centered world within which we live. Admittedly, the culture we have imposed on ourselves is for many of us the entirety of our environment. Without a walk in the woods, our habitat becomes the mall a mile away, the mall being built next to it, and the movies and media that we see there.

Although the influence of overwhelming consumerism and violent million-dollar movies on our way of seeing was not the general conclusion drawn from experiments in cortical plasticity, making the connection to cultural influence is only a short intuitive leap away. The kitten and monkey studies were done during particular developmental stages, and the experimenters rightly concluded that visual function is largely shaped during the early stages of development. But there were additional findings indicating that adult animals—cats and grown monkeys—also show changes in visual function and cortical structure due to experience. The obvious extension is the recognition of the degree to which perceptual habits and ways of seeing are continuously conditioned by culture, even as adults.

It is no surprise that much of how we perceive is shaped by ongoing experience, by the scripts we follow, by the voices of those around us, by cultural coding. We all share and reinforce the daily assumptions that constitute consensual reality through the language we use, the media we watch, the stories we tell, the science we value, and the methods we use to describe truth.

Although visual science is informed by physiology and physics, and is occasionally inspired by findings in evolutionary biology or anthropology, it is generally hidden within psychology departments. And although academic institutions are relatively invisible to most of us, the scientific definitions that arise there infiltrate our common consciousness, particularly in the realm of psychology. Because psychology is all about human experience, it is easily popularized. We all know something about it by virtue of being human, and we all have at least a passing interest in it. As a consequence, the heady esoterica exchanged by academic psychologists is all too often reflected in a body of common assumptions. Given this, it is worth repeating the wisdom of James Hillman: "We see what our ideas let us see." Nowhere is this more true than in our assumptions about vision. Our ideas about seeing filter and shape what we see.

Postmodern Perception

> The outline and colors are no longer distinct with one
> another. . . .
>
> —*Maurice Merleau-Ponty*

When I say "postmodern" I am referring to a form of shared thought that emerged during the latter half of the twentieth century. This form of thought appeared in cultural events, in the arts, and in attitudes about the way of the world. It represents a growing recognition of cultural relativity and, as an extension, an unwillingness to pass judgment on *anything*. Within a postmodern context, no single life choice or cultural custom is thought to be right or wrong. Any given behavior or choice is contextual and

cannot be judged outside the context that defines it. Richard Tarnas, in *The Passion of the Western Mind*, concludes that the postmodern mind is ambivalent, uncertain, caught in ambiguity. But Tarnas
also believes that this perspective represents an opportunity to
reframe the world, to restructure a worldview—one that incorpo-
rates our growing recognition of relativity. Although postmodernism is a complex mix of philosophical and political thought, I
wish to simply acknowledge this shift in consciousness as a movement toward contextual ways of perceiving reality. I see postmodernism in contrast to modern consciousness in which the
mechanistic viewpoint implied absolutes, the predictable and unchanging face of reality. Postmodernism represents the reemer-
gence of a relational way of seeing. Nothing is absolute or given in
a relative world. Postmodernism includes our developing consciousness of the relationships that constitute a woven, interdependent world.

By many accounts, we are well within the dawn of an ecological
age. From economic, political, and environmental perspectives,
there is a growing recognition of global and radical interdependence. As we turn our focus toward the relationships that constitute interdependent reality, what, then, does it mean to see well?
Does acuity still mean 20/20 vision measured with black-and-white
letters on a two-dimensional eye chart? Or might acuity come to
mean one's ability to "see into," to see beyond the sharp distinctions
between objects to a realm where the world is known to interpen-
etrate itself at every edge?

Hunting guides and trackers are known for their ability to see
animals that hunters fail to see. The difference between the
hunter's perception and the guide's is that hunters typically look
for game, for isolated animals, for the category "deer," whereas

guides attend to the landscape, the context. This relational perspective is oriented toward that which changes within a given context. Guides are attending to contrast and anomalies in the landscape, to a disturbance in the patterns. Hunters are caught in object identification, a way of seeing endemic to a modernist view, a culture wedded to materialism and reductionism, to objects and parts. There is no surprise in this. Guides, on the other hand, are looking for those distinctions that become apparent only from a relational perspective. They look for what shifts, for the edges that blur. Acuity in this sense is the ability to see change, that which is dynamic, and this can be seen only with an eye toward the entire field, the context.

I wonder if acuity, in an ecological age, is merely another word for having a refined sensitivity to that which is beautiful, that which has integrity and wholeness. For the Shipibo-Conibo, the term for "aesthetic" is also the term for "appropriate." To them, that which is aesthetic, what we in the West might call "beautiful," is really a sign, a form of guidance. Doesn't this way of looking make sense? Doesn't it make sense that what draws us by its beauty is somehow "appropriate" and thus offers us guidance? For the Dineh (the Navajo), the "Beauty Way" describes just such a guided path. The Beauty Way simultaneously describes a way of being, a way of viewing the world, and one's worldview. As a way of being, the Beauty Way is participatory and engaged. One's body is tuned to resonate, to respond to the subtleties of that which carries beauty. Traditionally, this included the everyday awareness of the four sacred mountains—Mount Taylor, Mount Blanca, the San Francisco Peaks, and Hesperus Peak—that encircle Navajo country and of the indwelling spirits that live there. It includes the belief that Spruce Mountain is the home of Changing Woman and the heart of the earth; that Mirage Mountain is the place where one goes for refuge. Following the Beauty Way means being immersed within a storied landscape guided by *hózhó*, the notion of a continuously changing order, a dynamic balance, a shifting harmony.[35]

Living within the Aboriginal Dreamtime is, I imagine, an analogous way of seeing the world. The Dreamtime is at once sensation and knowing, a seamless form of perception. It is both the profound observation of the physical landscape and a sophisticated form of knowing within the metaphysical realms. Paul Devereux, author of *Symbolic Landscapes*, says that the landscape is "the symbol of, and gateway to, the great unseen world."[36] He continues, "[The Aboriginals see a] profoundly symbolic language in topographical features," thus weaving together the visible and invisible realms. From this perspective, the powers of symbol and metaphor are built into one's way of seeing, and "sacred" is simply that which unfolds from such a view of the world.

If vision—and perception in general—is a form of translation between inner and outer geographies, then the question of vision, of how we see, becomes one of how we translate between these realms. What do we bring to the marriage between the sensible world and the realm of personal psyche? What is the quality of the relationship between the energetic flux, the light we absorb, and our own state of being and consciousness? Are we engaged with the subject of our gaze, or are we spectators—the bystander, content with passive observation? And what difference does it make, anyway? Why should the way we perceive matter to us? Should we be content with perceptual ruts, with our comfortable attachments to particular forms of reality? Could sacred ways of seeing change our behavior toward the worlds we inhabit?

"To see" is a relative process defined by a wide range of experience. It might mean the perception of surface qualities, of mere appearances; it might be the pure perception of phenomena, a direct, unmediated way of "just seeing," a "Zen way." There is also the embodied quality of Merleau-Ponty's sight, the intuitive and practiced perception of Goethe, or the perceptual communion of David Abram. One can define seeing as the acute awareness of form, of negative space, or of events. A careful definition of "to see" might also include a genuine perception of depth, revealing our embeddedness in multidimensional constellations of reality. Rudolph Arnheim, author of *Art and Visual Perception*, defines eyesight as insight, reiterating a historical and indigenous view of the seamlessness between physical and psychological realms. Similarly, there is William Blake's famous "double vision" in which the mythic is rendered in the mundane. Perhaps ultimately, acute perception arises from a form of presence so deeply woven with incoming signals that we lose the distinction between seer and seen. Perhaps this is what Empedocles, looking at the world through something akin to "love eyes," meant by "to see." From the perspective of modern modes of perception, perhaps Empedocles' sight sets a standard for postmodern acuity.

These ways of seeing overlap and tumble into one another until, at one end of a spectrum defining "to see," Western notions of apperception and shamanic forms of "vision weaving" become indistinguishable. This mixing of sensory and symbolic data depends on cultivating a participating mind. In the West, this type of presence and perception can be seen historically in the mystic and medieval mind and in what Devereux calls, in *Symbolic Landscapes*, "neolithic consciousness," a state of awareness that preceded the emergence of the modern mind.[37] This state of awareness is analogous to living in the Dreamtime, to walking in the Way of Beauty. It is a participatory state of consciousness, the doors of perception banging on their hinges. For those of us buried in the conditioning and flat material seduction of our era, practicing participatory forms of consciousness, or developing an "organ of perception," might help us to develop what environmental educator Mitch Thomashow calls "biospheric perception." This way of seeing is grounded in science—infused by what we know of reality from biology and geology, geophysics and astronomy—and yet is shaped by imagination and poetry. Thomashow writes that biospheric perception is "the place where perception meets analysis head on through expanded sensory awareness."[38] As a mode of perception, this way of apprehending the world moves easily across scales, embracing local and global ecologies in a single, penetrating observation. It both observes the biogeochemical flows in the streambed at one's feet and helps us to recognize the patterns of our collective behavior in the stream's flooded and eroded banks. In the flooding and silting at our feet we see the upstream clearing of forests and the resultant and rapid erosion.

One way of understanding the relativity of "to see" is by thinking of it as stretched along a continuum. This continuum ranges between the careful observation of the material world and a sophisticated understanding of the inner landscape. Our sensitivity to contrast, to the variations in light intensity and wavelength, can be found on one end of the spectrum. At that end of the spectrum, our vision is literal and measurable. There we can observe the sensory qualities of texture, hue, and value. At the other end of the spectrum is our ability to grasp the gestalt offered by metaphor and myth. At this end we are able to perceive signs and symbols, the meaning of the resonance we feel with certain people, places, and things. Along the length of the entire spectrum is the fact of our embeddedness within particular cultural categories, the fact that our whole way of seeing is colored, and sometimes veiled, by cultural prescription.

The image of such a continuum helps us grasp the seamlessness between the world of appearances and the invisible realm, those realities which are yet to be revealed. However, the image of a single, two-dimensional continuum cannot also convey the sliding energetic flux between myself as viewer and the viewed, the fluctuating and shared emanations. And it is this dynamic thread between the seer and the seen that helps me to find beauty and to fully understand why David Abram describes perception as communion, as intercourse. The recognition of this thread helps to satiate the hunger aroused by the taste of "participating consciousness" and by the visceral feel and flavor of multi-modal sensory awareness. And it begins to ease my urgency, for with a recognition of perception as a truly participatory act, I also recognize some hope of satiating my desire to respond, to be of use in a world that needs vision and visionaries.

Larger than that conveyed by a spectrum describing the relationship between inner and outer realms is a view of perception that integrates the flux and seamlessness of relational ways of seeing. Sight in this larger context becomes sensibility, the simultaneous integration of the multidimensional sensible world with imagination, history, and nuance—our own multiple dimensions. Despite the dominance of vision in Western experience and consciousness, our language clearly represents constellations of sensation and previous experience. We talk about our sense of time, inseparable from place; our sense of place, inseparable from time. We express a sense of gratitude as the sun drops from view in a crimson-colored sky. We comment on what we sense from the inside of ourselves, recognizing the ways our bodies shift in response to the energies entering us through our sensory channels, infiltrating and gathering into forms of meaning and embodiment. Sensibilities are constellations of sensory moments—of seeing, tasting, touching, smelling, and feeling—that pulsate through the body and awaken the soul.

When we embrace this experiential mix, the world around us grows in meaning and influence. We perceive a convergence of forces—light falling into our eyes, energies drawn down and absorbed into the body, mixing our history with the wide existence of the world. We recognize the color and scent, the voice of the world filtering through us in light reflected off rosy-red boulders—salmon and peach, the warm colors descending into our root chakra,[39] finding the right correspondence, merging with like energies. This is the long, slow-wave red light of the visible spectrum running through the fibers of the body, descending to the root, to blood. We pulse with the color red.

"To see" includes listening and hearing, touching and feeling, sniffing, licking, and tasting. These are the ways we sense our world. The salty air touches tongue, its taste circulating through the body. The forest rays into our tissues, thumping the chest, knocking at the heart chakra[40] as we walk in green. To see is visceral, symbolic, a path to enlightenment, a choice. Unfortunately, those of us conditioned by Western values tend to bring little awareness to the exchange between ourselves and our surroundings. We seldom consider the degree to which belief and consciousness shift and determine our particular way of seeing, or the way in which perception is a shared flux of fire, emanating from both sunlight and soul. Perception is intercourse, a tantric affair with the folds and flushes of the landscape.

To take vision for granted, to be thoughtless about the variety, miracle, and magic of seeing, is to limit one's self to a play-by-play objectification of reality. But to envision the act of seeing as the marriage between the viewer and viewed is to be woven into the fabric of a shifting field of light, of energy, beauty and all that one may lay eyes upon. It is to recognize that, as in all marriages, there are a thousand ways to honor the union. And as in any intimate relationship, what we bring to the exchange determines the quality of the experience.

Dreamtime

> In the Dreamtime time does not run out. . . .
> —*Paul Devereux*

I awaken in the very early morning, knowing that my dreams have been trying to pry open my understanding, to teach me something. But like the early dawn light that barely reveals the outline of tall trees outside my window, the teachings are there and not there. I grasp only hints of both light and lessons. I dreamt of symbols slipping and fading, of a fading presence, of a loss that we are experiencing now, together. I wish I had the strength of mind to call it back, whatever this loss is. Slowly, as if surfacing from beneath a diaphanous veil, my dream comes back to me. I dreamt the loss of the Dreamtime itself. Symbols of the Dreamtime slipped and disappeared over the curved horizon of my dreamscape, but I could not grasp them. A raven calls loudly, punctuating the morning as if to scold me, as if to mark my accountability for such a slippage of mind.

I weep all morning, so great is my sense of loss.

I cry for the waking dream, for the state of consciousness that enables Aborigines to see a landscape laden with the meaning and magic so missing from my modern mind. Contrary to the thick veils in my dreaming, Aboriginal Dreamtime arises from profound, awakened sensory awareness. Traditional Aboriginal children grow up in a world that demands acuity, a world that will feed them or not, depending on how well they see and listen. It is no surprise that by three to five years of age, Aboriginal children are able to feed themselves by hunting small animals. By adolescence they are able to identify the individual footprints of two hundred, often three hundred, clan members.[41]

Not only does the Dreamtime demand profound acuity, it is synaesthetic: sensations merge and mix, becoming undifferentiated. The experience of synaesthesia—tasting the scent of root, for example—is intensified, like the Dreamtime itself, with profound sensory awareness. In Western science, the apparently slippery ability to mix sensations, to taste a color or scent, has been linked to "having a good memory." But anybody knows that a good memory originates with sharpened awareness. We remember the moments when our attention is fully drawn to a particular color, form, or whisper, to the sounds we hear and the scent of a place entering into us—the moments when our focus sharpens. Then, fed by fine memory, we shift sensations into a new arrangement. We translate sensations across the senses, loosely, with confidence and ease, relinquishing boundaries. A "fluid ambiguity" surfaces: I *feel* the familiar cottonwood leaves I *watch* in the wind, feel them as though they were fluttering against my thigh. Such is synaesthesia; my sensations are mixed and nothing is apart.

The revelation born of seeing and feeling in one seamless, synaesthetic moment is that nothing in this world is unrelated, not a single thing. The boundaries blend as if a heightened awareness has begun to reveal permeable membranes between the things of the world, and between ourselves and the world. Distinctions blur and relationships emerge, shimmering and shining like light on water. Where does water end and light begin, anyway? We see the glitter, the relationship itself. Each thing of the world becomes more than it once appeared to be. And all of it quivers, dense with possibility.

But in the Dreamtime, what appears as possibility *is*, the ancestors *are there*, the Dreamtime beings *are* in the bushes, just there, beyond the light, teasing and telling you where you really are, "where you're at" in the truest sense. The world signifies, beckons, speaks, as if to say, "Now, see this." Look, see the image emerging, arising between the named and known things of the world. Oh, but between the names, beyond the world of appearances, I grow confused. How can I *feel* the leaves I *watch*?

Synaesthesia is also mundane. It happens when language, an auditory signal, is translated out of itself and into images. We see pictures in our mind's eye, the images informing us to a much greater extent than any string of words. The images are sculpted and embellished with memory, the sounds and scents that have taught us the world. Synaesthesia becomes the waking dream, an awakened stream of images arising from all the senses. From my Western perspective, I am certain that these synaesthetic constellations are the substrates of sensibility. One's sensibilities are, after all, textured by time, scented by season, and responsive to incoming influences in every moment. My sensibilities are constellations of me, the present and the past, all at once, "a kind of time out of time." And they "spell" us; I am cast under the spell of my senses arranging themselves into me-in-a-(multidimensional)-moment, singular and larger than myself and my linguistic abilities. I have no words while dancing in wind and weather under a setting sun, perhaps a few tears and a desire to throw myself down onto sand, wrapping arms and legs around the great curvature of the earth, holding this big ball of beauty against me as if I can't get close enough. Yes, this is beginning to make sense. *Making sense*, as David Abram says, is simply the senses vibrating and resonating. *Making sense* happens when the wind—rising up and stirring, as if from sleep—passes through leaves, at once brushing against my thigh and filling my breath, stirring me and constellating in the passage of another (oh-so-sensuous) day.

This is the way in which spells are cast upon us, the wind casting a spell on the wholeness of my senses, shaping a seamless reality into my body. In *The Spell of the Sensuous*, Abram reminds us of this power to be "spelled." We are spelled by the world, shaped by the gathering together of our senses with the sensible world, gathered in the act of divining (or not) our place, this ground.

The Aborigines say, "White men have lost their dreaming." Aboriginal dreaming is tuned for receiving gestures, signs, and potencies directly from the pulsating voice of the earth, from the pervading echoes of the ancestors. The Dreamtime, I think, is the consciousness of the land, arising from both visible and invisible realms. It must be the soul, the psyche, the breath of the world (of us) informing in every moment, whispering to us (even as we whisper), telling us where we are (telling the earth we are here), telling us in the voices of contour and constellation, in languages created within wide webs of intelligence, speaking the dream, arising within the landscapes of all (of us). We answer back (they hear us), muted or pleased, we choose to be where we are.

I dream again. There are signals streaming through the pores of my antlers, an unceasing flow of low-frequency vibrations. I am satiated. When I awaken, the morning is filled with tears of (radical) joy.

Chapter 2

Numb and Not-Noticing

How the Modern Eye Sees

Y ou have to work with your heart and your vision,

what you hear and feel.

If you don't use the eye anymore, it fade out.

If you don't use the ear, it fade out.

If you don't use the heart, it fade out.

That's why you see so many blind people. . . .

—Bobbie Billie, Seminole traditionalist

Real-World Circumstances

> We know not where we are.
> —*Henry David Thoreau*

Crickets call to the east. A chopper ratchets a mile to the west. I sit in the middle, my left ear seduced by the soft cadence, the ever-shifting song of crickets in spring. My right ear is hollowed out, hard, both braced against and invaded by the clipped din of machinery. I am entered without consent. If perception is intercourse with the sensible world, then my right side is being raped by noise.

I am beginning to cry. I have felt the breath and nudge of the Dreamtime and know that it is just beyond my threshold of perception, just beyond my reach, just a slip of consciousness away. I long for my serpentine thirst to be quenched by the dreaming, long for the look and feel of ultimate belonging and the sensuous play of being embedded, in bed with the world, dug in and dirty. But the phone rings, my endless list of things to do nags, haunts, and fills my consciousness. I too perceive the invisibles. In this case, they are mostly petty preoccupations—the trip I must make to Safeway, the phone calls I must return, the mail piling up—and the fact of my father, growing old, alone, 3,000 miles away. A phone call to him does not appear on my list. I feel such sadness as daily obligations fill my badgered view. I go blind in order to forget.

The daily demands of our lives cause us to narrow our field of vision, shaping and minimizing our view to match a preoccupation with phones and faxes or a long list of tasks that are never complete. Then in unconscious defense against the onslaught of modern business as usual, we further minimize the sensations we receive with self-inflicted doses of numbing. Most of us, I dare say, are numb to varying degrees, and for good reasons. In the late twentieth century, we can see, taste, touch, and feel the degradation of our planet and, more immediately, of our own neighborhoods. We are continuously confronted by the noise and nonsense of the media, the superficiality of politics and plastic, and the tasteless seduction of Wal-Marts and K-Marts. For a sensitive being, or for human beings with highly evolved sensory systems, all of this is painful. Perhaps the most painful modern reality is knowing that each of these "lowest-price-ever" outlets carries the karma of environmental destruction in the form of relentless extraction of planetary resources. No wonder we are no longer sensitive to the pulses and whispers of sensation. Sensitivity hurts. Rather, we comfortably collude with modernity in a sleepy form of denial. While coveting another plastic gadget in another plastic bag, we conveniently forget the implications of our unwitting participation, and the shame that goes along with it.

Living in an era of ubiquitous environmental degradation is depressing to sensitive organic systems. Humans are no exception. Like many of us, I turn down the volume to protect myself. I become psychically numb—a classic psychological defense against that which hurts. This state of being is what James Hillman calls "anesthesia." Anesthetized, we no longer gasp in sudden wonder, inspire or become inspired as the beauty of the world enters us, for we are artificially numbed—as if shot through with drugs to induce non-awareness. David Abram calls this state "collective myopia," implying that we see little beyond our comfortable and constrained personal environments; we lack depth perception. Theologian Thomas Berry calls us the "autistic generation," suggesting a massive defense against an overwhelming assault of uncomfortable sensations and implying our collective inability to respond. Naturalist and writer Terry Tempest Williams asks: "What is it that we are not seeing? What is it that we are not hearing?" Referring to technology, the world of our own invention, Jerry Mander suggests that we are "essentially living *inside our own minds . . .* essentially co-evolving with ourselves in a weird kind of intraspecies incest," entrenched in a system of perceptions that make us blind and passive.[1] Chellis Glendinning, author of *My Name is Chellis and I'm in Recovery from Western Civilization,* puts this more succinctly: "Virtual reality," she says, "is a sure fire perceptual escape."

The critics of our time are numerous. But we already know the message. Despite what may be a collective form of myopia, we cannot entirely escape the recognition of having eyes trained on TV or of drinking and drugging in sleepy defense. But do we see the full extent of our denial? We continue to busy ourselves with a multitude of distractions. Mostly, we buy more things, as if the natural world does not despair by our endless taking. Are we blind?

While driving through the desert just north of Phoenix, I listen to Natalie Merchant as she sings, "Have I been blind . . . / hypnotized, mesmerized / by what my eyes have seen." I see white plastic, black plastic, and clear plastic bags and bottles strewn over the roadside desert. I am just north of the penitentiary, just below Pinnacle Peak. I envision penitentiaries overfilled across the country, pinnacles of modern civilization. I pass a bus full of prisoners, followed by the prison truck, capped with red, white, and blue cop lights and marked with emblems of authority, as if the driver carries the U.S. official answer to the question of crime. I pass a freight truck hauling hundreds of bags of "plastic cement." I enter Phoenix—"a carnival of sights to see" veiled in smog and bill-

boards. Next I pass Deer Valley Road knowing that the deer are
long, long gone, having been chased by earth-moving machines
that gouge and dig the desert at the rate of several acres per hour.
I pass under four freeways arcing into four nondescript directions,
with two unfinished cement monoliths yet hanging in the air. I
drive my car, watch a box tumble from a truck, spilling unknown
white powder onto the freeway. This is both landscape and an all-
American true story. In response, I close down, protecting a sensitivity that has evolved over millennia. It hurts to be awake in such
a carnival, one determined and blindly delivered by the controlling interests of commerce, capitalism, and more.

As I write this, we serve several million doses of Prozac each
day within the United States alone. In 1990, 650,000 new prescriptions of Prozac were written each month. In one year's time,
this translates into 7.8 million new prescriptions. At this rate and
over a decade, 30 percent of the population of the United States
will have received 78 million prescriptions of Prozac.[2] Unfortunately, Prozac is only one of a dozen or so such modern drugs in
common use. The cascading use of antidepressants certainly suggests that depression is more than an individual affair. Rather than
reflecting personal pathology, it may reflect—from the perspective of our very sensitive sensory systems—the intolerable collective conditions of modern, industrialized life. But as the diehard
individuals we are, we do our very best to keep up with the Jone-
ses, while lining up secret vials in the medicine cabinet. Given
modern conditions, our true desire might be to simply lie down,
stare at the wall, retreat, and reflect on the state of our sad souls.
The symptoms of clinical depression include not sleeping, not eating, or mindlessly eating more. We don't think very well when depressed; decisions are hard. We experience little interest and little

pleasure, and a notable loss of meaning. We radically dampen down the senses, slow the fire, and don't notice much. Or if we do, it is within comfortable confines, a narrowed visual field, near and safe. We stay in bed. We hide under the covers. Heaven forbid the light of day.

Although by now the twenty-first-century mind should be well aware of the interdependent links between earth degradation and widespread depression, our behavior has not yet caught up to our awareness. In a fog, in a hurry, and apparently in a hungry search for purpose, we continue to consume dwindling resources in the form of plastic paraphernalia, marketplace fashion, and endless other artificial and personal prescriptions for a life. This is psychic numbing, denial, alienation, and depression wrapped into a single package, a single pattern of unintentional and unconscious misbehavior—a pattern we might call overconsumption or addiction. Most insidious, this effect is cumulative, leaving us increasingly blind, autistic, and misguided in a self-perpetuating geography of human invention, a shiny and seductive world of mirrors. Saddest of all, we have forgotten that we have forgotten. We are too far from the richness of sensory, sensual experience to remember, much less taste, what it means to be fully awakened and engaged. We no longer know what a depth of sensory awareness is. The outcome of our forgetting is a sleepy, comfortable denial of our loss.

You may ask, with perhaps a hint of denial, what have we lost? We have lost the touch and taste of the world, the hint of color in light, the subtle swing in a bird call, the gasping and crying and kissing because a brilliant, sensational sunset reaches into our eyes and gets under our skin. In our forgetful state, we don't bother to look. The rhythms and patterns of the earth slip away. There is less that intrigues, that captures our attention. We see and feel even less, missing the signs and symbols offered by the world, and especially by the natural world. We lose the sensations and feelings of being part of, somehow buried within fields of sentient and even sexy relations. We become lonely, hungering for a sense of belonging, for true home, that too having slid from view. We lose the whole picture, and a sense of wholeness. We drink and drug in a mad search for satiation. It makes us sick, our souls so unsatisfied. Ultimately, we lose our health, our own wholeness.

Psychic numbing is prevalent and insidious. Numbness and the oft-quoted "loss of meaning" describe the psycho-spiritual conditions of our times. And although we all know the feel of it, we seldom recognize the depth or degree of our denial. Because the nature of psychic numbing is to distance and diminish, it is difficult to detect in our selves. We are able to see it, however, in contrast to the experience of others who are still vibrant with light and life streaming through them. We hear the depression of our own sensory selves in contrast to the rich stories of skilled trackers, or in contrast to the thread of wildness still running through the stories of Alaskan fishermen. We may also reveal our denial by radically changing the conditions of our lives, by entering into entirely new landscapes, the spicy foreign lands that demand our attention and shake up our worldview. And we can recognize our numbed defense by virtue of the mystical moments when our habitual awareness is pierced through, when long shafts of late-afternoon light split open our crusty cynicism, our habitual despair. But these moments are fleeting, as if stripped from consciousness in the unfamiliar moment of their occurrence. There's no doubt that Western conditioning has made the mystical moment into a form of deviance.[3]

True Sources

> A great deal of our ideology is hallucinatory. One of the most
> important things for us to do is get back to our senses, in the
> palpable sense of responding directly to physical nature. That's
> why I'm so interested in the Enlightenment. It was when philos-
> ophy went off on this curious tangent where moral considera-
> tion became the sole possession of 'rational beings,' that is,
> eighteenth century Europeans in white wigs.
>
> —*Christopher Manes*

Traditional Australian Aborigines spend the greater part of daily
life dancing, chanting, and inviting trance states of consciousness.
They cultivate the Dreamtime—falling under what we might call
a kind of synaesthetic spell—with both intensified sensory awareness and the mixing and mingling of sensations. We know little of
such states of being. Anthropology, travel, and the media have
served us well by revealing the possibility of diverse ways of being
and perceiving. But this awareness has engendered a kind of postmodern paradox. Can we access sensory states once we know they
are possible? Can we simply "get there" by knowing such states exist? Through my modern eye, the tease of another reality translates into a kind of postmodern paralysis. Like many of us, I go
hunting and gathering at the mall.

This is what Chellis Glendinning refers to as the outcome of
the "original trauma," the fact of our modern alienation from the
natural world, from the places of our very long human evolution.
With the disappearance of hunting and gathering, our survival no

longer depends on our sensitivity to the flicker of movement in brush or the scent of tasty plants in the next clearing. Without our senses being called out for the sake of survival, we do not experience our full sensory capacities, our full humanness. And so we do not fully know ourselves. Instead, we muse through a Victoria's Secret catalog, acting as if we might find satisfaction by hunting through flat, one-dimensional glossy pages.

For writers like Glendinning, the primary source of our estrangement from the nonhuman world is agriculture, our domestication of the land, and the consequent separation of ourselves from daily and deep engagement with the wild. Some ten thousand years ago, we began building fences to protect our crops from the wild. Fencing and owning our own ground could be seen as both a brilliant idea and a form of domination over anything that could be cultivated. But the joke is on us. We have fenced ourselves from the experience of wildness, from the core of our own nature, and from a deep-rooted resonance with the ultimate sources of creation.

The language we commonly use further shapes and shifts our perceptual sphere. We speak of inert, dead matter, objects without life and with which (notice: not "with whom") we cannot interact. Resources are spoken of as "for our use" and fragments of wildness are "vacant lots." But it is not simply the daily discourse that alienates—it is also the fact of writing. I am convinced by David Abram that by writing down our experience, we lost the genuine touch and feel of the earth. In the act of abstracting and placing our experience on two-dimensional pages, we cleaved those experiences from the land—the very source of sound and of our desire to communicate. Our experiences, now encapsulated in written stories, became separated from the ground of their inception. We read our stories quietly and independently, without the forms of engagement that shift and change a story, without adaptations to each place and the very moment of the telling. The story became static. Our sensory engagement with the story—now stripped of wind whistling or sun shining—became minimized, the sensory body forgotten.

The roots of our alienation from the sensible world, from both the landscape and our own rich experience of it, are many, interwoven and tangled below the surface of the modern mind. Most fundamentally, our separation from the natural world has its source in amplified forms of dualistic thought. In the West, dualism is the largely unconscious tendency to divide reality down the middle, into either/or categories. We want to know if something is good or bad, right or wrong, moral or immoral. Reality is cast in black-and-white terms, few distinctions are made, and there seems to be a factual answer for everything.

Within dualistic forms of consciousness, distinctions tend toward divisive perspectives so that we see this *or* that, but not *both*. We perceive things as existing not only at opposite ends of a spectrum but on radically different spectra, in different realms. We see good or bad behavior, right or wrong answers. This disavows the relational aspect of reality and renders our experience absolutist in flavor, rigid in feel, and without texture or context. The world becomes less engaging and more easily objectified. The object-other is diminished and dismissed.

Dualism is a defensive response to living in a world of continuous change, an interdependent reality. Ripple effects continually run through the system and nothing stays the same. Because of this inevitable and ongoing change, fact becomes illusive and we simply cannot know truth in absolute terms. Not knowing for sure is disconcerting to the Western mind. It is discomforting and can even be threatening. We build barriers, amplifying our sense of difference between ourselves and whomever the other may be. The cumulative and corrosive element of dualistic consciousness seems to manifest as oblivion. We are concerned about subject (read: ourselves) and not object; object becomes little more than something to satisfy ourselves, and with a now one-dimensional interest, the experience of depth becomes irrelevant. We end up not only alienated but also living as "surface dwellers." I see strip malls, long, lined up, and flat. They are impenetrable facades littering urban America and further setting the stage for all sorts of "not-noticing." Separation and the consequent dualism, recapitulated throughout millennia—perhaps since the advent of agriculture—are now squarely in our visual field, in our daily experience.

Dualism is also squarely in our lineage. A long legacy of dualistic thought has been etched into our Western psyche. Cartesian coordinates map relationships onto two-dimensional, flat surfaces that describe little more than causality. What we refer to as Cartesian dualism honors Descartes's proclamation, in the early seventeenth century, that mind and body live in separate realms. The roots of such two-dimensional thinking are even older than Descartes. It was the Neoplatonists of the third and fourth centuries who believed that the root of all evil resided in matter, and particularly in the flesh. They, of course, followed in the footsteps of Plato, whose complex philosophy had profound implications for Western consciousness. Plato's dichotomization between pure thought and material reality became the apparent source of the infamous mind-body problem, the mind-body *dichotomy*, in which the transcendent capacity of the mind is cleaved from the physical body, the material world. The body became reduced to a mechanistic collection of parts, all to be conveniently mapped, numbered, and further separated from the Gods. The mind-body problem became a classic case of dualism.

Although the events and thinkers that influenced the evolution of dualistic consciousness are far too many to name, a few stand out as having particular power. In the shadow of black magic, devil worship, economic depression, and corruption by the church—in essence, "massive cultural decay"—the Black Plague mysteriously and horrifically swept through fourteenth-century Europe. *What was it* that, within three years, left a third or more of Europe's population dead?[4] *What was it* that swelled the armpits, the groin, became blisters, boils, oozed blood and pus, and then became mysterious black blotches? What was the fever and blood-spitting stench that accompanied the whole mysterious mess? Victims died quickly, at most within five days of the first symptoms, and sometimes in the space between going to bed well and never waking the next morning. The absolute inability to discern a cause gave the plague a supernatural, sinister quality. Even the pope called it God's wrath, leaving his many followers with an expanded sense of guilt. Because, in the eyes of the church, mere failure to fast or to attend mass was sinful, the plague's legacy included a simmering discontent with the excessive and confusing demands of the church and a silent distrust of God. No surprise, then, that the plague also left the medieval mind with a hunger for certainty, a desperate need to know in absolute, black-or-white terms why the world was the way it was. This hunger for certainty paved the way for the unfolding of scientific rationalism, which was just around the corner.[5]

The burning of witches during the sixteenth and seventeenth centuries furthered the rationalistic enterprise and left deep scars on the Western psyche. Some estimate that 9 million women, children, and men were burned over six generations for the practice of witchcraft, for divination and healing unmediated by the church or state. The church was, by now, clearly not in control of the common person—and was apparently very threatened. One church official alone, Archbishop Como, ensured the burning of one thousand witches within a single year. Even Joan of Arc was burned at the stake for receiving direct guidance from a favorite tree. In the end, the burnings—the sixteenth-century holocaust—effectively silenced sensual knowledge, the intuitive knowing derived from an intimate relationship with plants, animals and places, with the natural powers.

The major inventions of the early Renaissance—gunpowder, the magnetic compass, the mechanical clock, and the printing press—all had a major impact on identity, on how people came to view themselves; identity became threaded with individualism, nationalism, and a heady new sense of invincibility. Gunpowder contributed to the demise of feudalism and the more visible power of the nation-state. The compass made crossing oceans relatively easy and, along with gunpowder, offered the entire planet for the taking. The clock freed human activity from daily dependence on (and intimacy with) the rhythms of nature. It led to new notions of linearity, progression, and progress and served as a model for the emerging mechanistic worldview. The printing press produced a tremendous increase in learning, ending the monopoly on textual learning held by the church. It stimulated reading on a massive scale, an altogether new genre of private experience, and allowed for the rapid dissemination of radical ideas. In the Western mind, these influences translated into a newfound heroic ideal. Sailors, soldiers, merchants, and investors were called to new lands, to entirely new frontiers. The world was to be explored, discovered, and conquered.

With the West's rapid expansion on all fronts—culturally, economically, geographically, and politically—and with the consequent recognition of entirely other ways of life, a skeptical relativism was introduced into the Western mind. European traditions and assumptions were no longer the whole story. In contrast to the explosive possibilities represented by the widespread exploration of the early Renaissance, European assumptions were more akin to constraints on personhood. And so there emerged the infamous "Renaissance man." The Renaissance man was skeptical of orthodoxy and the dogmatism of the church, no longer dependent on an omnipotent God, committed to a greater future, rebellious of authority, and newly assured of his capacity to control nature.[6] According to Richard Tarnas, in *The Passion of the Western Mind: Understanding the Ideas that Have Shaped Our World View*, "The medieval Christian ideal in which personal identity was largely absorbed in the collective Christian body of souls faded in favor of the more pagan heroic mode—the individual man as adventurer, genius, rebel. Realization of the protean self was best achieved not through saintly withdrawal from the world, but through a life of strenuous action in the service of the city-state, in scholarly and artistic activity, in commercial enterprise and social intercourse."[7] Together the unfolding of events and inventions ushered in still more heady self-confidence, still more hunger for certainty, and the newly established notion of progress. It comes as no surprise, then, that science became the new faith.

Science promised definite answers. It was empirical and rational and appealed to both common sense and concrete reality—it described the width and depth, the heft, the measurement of the world. Facts were verifiable and reproducible; theories could be proven and could replace the dogmatic revelation previously imposed by the church. But the most seductive aspect of science was that it ennobled the mind, playing on the capacities of human comprehension and reason, amplifying individual competence, and delivering inventions and technology. Science suggested, even promised its followers, a kind of liberation. Individuals could attain certain knowledge through their own reason and observation.

All of this further ensured the emergence of rational, dualistic thought. It paved the way for the revolution that marked the full-fledged arrival of science and, with it, the certain control of nature. Francis Bacon, often named the father of reductionism and of scientific methodology, is famous for his treatise on the "vexing of nature"—the discovery of truth by putting nature in a position in which it was forced to yield answers. "Vex nature, disturb it, alter it, anything—but do not leave it alone. Then and only then, will you know it."[8] And knowing it meant being able to use it. For Bacon, knowledge was power and true knowledge could only be measured by how it could be used. A mechanical invention, from Bacon's point of view, was infinitely more useful and more indicative of intelligence than was poetry or spiritual wisdom. So influential was Bacon that by the seventeenth century, and under an avalanche of influences, the modern man had turned his eye away from the uncertainty and wild wisdom of nature.

There is no doubt that science has provided us with much comfort, convenience, and a tremendous depth of understanding about the nature of physical reality. But the wedding of Cartesian rationalism and Baconian empiricism enshrined science, and as a consequence, sensual knowledge slid still further from view. The environment surrounding the enterprise of science became little more than pesky conditions to control. The feelings and sensations accompanying any observation became labeled as subjectivity—needless to say, a big problem for rationalism. And because science so readily translates into practical and technological applications, it is now intimately woven into every moment of every modern day. Science is behind our radios and clocks, our mixers, blenders, TVs, and the ubiquitous computers we are so utterly dependent upon. The continual presence of science in our lives, in the liminal realm between the seen and unseen, accumulates exponentially within our consciousness. The price we pay for such convenience is the costly dismissal of magic and mystery, of unmediated, direct experience and the sensuous life of the body.

Although this is by no means a complete tour through Western history, it provides a hint of its potent influence on the modern mind and on the modern eye. The particular themes I have emphasized still shape our vision. Granted, it is difficult to get a perspective on the combined forces of historical events and the conditions of contemporary life. Like fish in water, we are so immersed within Western monoculture that we do not fully perceive its nature, including an easy recognition of its conditioning powers. But the historical record reveals a conditioned thirst for secure knowledge, a belief in individual agency, a commitment to progress, and the habit of control. These themes have conspired in further conditioning a divisive frame of mind from which we peer into the world, from which we filter and make the world into a two- or three-dimensional fact.

The lens through which the Western mind observes reality can be characterized as basically unexamined and dualistic. We tend toward the crispness of fact rather than the uncertainty of mystery; we look for the right answer and do not contemplate the possibilities suggested by the many shades of gray between right and wrong or by the speckled interpenetration of subject and object. Looked at from the conditioned perspective of our dominant culture, the dreaminess of the imagination, the erratic qualities of the ever-responding body, and the irritating, nonspecific nature of intuition are all problematic. They are generally not considered to be legitimate ways of knowing and are thus not consulted in the context of any serious question or consideration. Artists, we presume, live on some faulty edge of reality, poets are essentially unrealistic, sensualists are not to be taken seriously, and, by the way, can't you make a little more money? Even success is quantified, counted, added, measured in absolute terms.

Despite my critique of dualism, it is worth remembering that it is simply an extreme form of a natural reality. The ground from which dualism springs—which is to say, difference—is inherent in any organic reality. Life is truly "all relative." It is the differences between things that give each thing its unique identity. This, of course, sets the stage for relationship to occur. And clearly, without more than one thing, there's no other thing to relate to. It is also true that it is only by way of difference or contrast that we perceive anything in the first place. Polar bears disappear against a background of snow, and the moon, pale in contrast to a daylit sky, is often overlooked. Without a difference in the visual field, neurons in the retina fall silent and even a highly saturated field of color fades to a dull gray in a matter of seconds. Differences are both clearly essential for perception and inherent in any relationship. Every out-breath requires an in-breath; male and female are fundamental to creation; and what's good (or bad) anyway? The answer is, "It's relative."

Huston Smith, author of perhaps the best contemporary text
on world religions, tells a marvelous Taoist story to exemplify the
Taoist view of relativity. It's about a farmer whose horse ran away:

> His neighbor commiserated, only to be told, "Who knows what's
> good or bad?" It was true, for the next day the horse returned,
> bringing with it a drove of wild horses it had befriended. The
> neighbor appeared, this time with congratulations for the wind-
> fall. He received the same response: "Who knows what is good or
> bad?" Again this proved true, for the next day the farmer's son
> tried to mount one of the wild horses and fell, breaking his leg.
> More commiseration from the neighbor, which elicited the ques-
> tion: "Who knows what is good or bad?" And for a fourth time
> the farmer's point prevailed, for the following day soldiers came
> by commandeering for the army, and the son was exempted be-
> cause of his injury.[9]

The Taoists greatly simplify and make visible the paradox of
relativity. The polarities are imaged as black and white, *yin* and
yang principles interpenetrating one another in a complementary
and balanced fusion. According to Huston Smith, they are "not
flatly opposed."[10] In the traditional form, yin and yang are encircled by a single line of embrace. Unfortunately, in our know-it-all
culture, paradox makes us uncomfortable. For the Western mind,
perhaps the more comfortable question regarding the nature of
dualism is one of degree. Have we amplified the distinctions inherent in any relationship—and in the relational world—to such a
degree that all we see are big differences? In the necessary act of
making distinctions, have we gone separatist, or even fundamentalist? Could rationalism, reductionism, mechanism, imperialism,

capitalism, individualism, and modern materialism have colluded in creating *extreme* separatism? Are we, then, alienated from one another and the natural world? Is this why we don't see and/or feel? And is this the root cause of those 78 million prescriptions of Prozac?

Disembodied and Mindless

> Our civilized distrust of the senses and of the body engenders a metaphysical detachment from the sensible world, fosters the illusion that we ourselves are not a part of the world we study, that we can objectively stand apart from that world, as spectators, and can thus determine its workings from outside.
>
> —*David Abram*

> If I'm not who you say I am, then you are not who you think you are.
>
> —*James Baldwin*

The historical and contemporary influences that have conditioned our consciousness have also influenced our Western bodies. Those of us with a Westernized mentality have been raised in a kind of collective training ground for being defended, for seeing ourselves as "separate from." The consequent loss of sensual knowledge is a loss of the recognition, intuition, resonance, and pleasure that come through the unmediated, unprotected body. Wilhelm Reich, whose many critiques of the Western treatment of the body were banned and burned, said, in no uncertain terms, "Our cultural history is encoded in our bodies."[11]

In somewhat confused contemporary terms we concern ourselves with the "boundaries" between ourselves and the rest of the
world. Despite our confusion, we seem certain that these boundaries must be "good." In my own case, this meant being the "good
girl," the one who grew up nonsensually—at least with reference
to the codified world of human relations. In a culture laden with
puritanical notions, being good, however, readily translates into
a denial of the body itself—the denial of hormones flooding
through the body, for example, and of the sensations and desires
that come with such floods. In the language of the body, being
"good" easily denies more than merely human relations. We also
deny our profound coexistence with the world. We deny our inherently erotic bodies, made of this earth and in intimate relationship with this place, this moment, this ground upon which we
stand and from which we arise. This denies us the experience of
our own true, embodied selves. My bounded, separatist self denies
the flood of light entering me in a sunny, sandy wash, the touch
and taste of a warm wisp of air slipping across my body, and the
feel of silken sand underfoot.

Admittedly, there are very good reasons for the development of
distinct boundaries. What we generally consider to be "good
boundaries" are frequently necessary in the world we now live in.[12]
But given our historical roots, the sad art of disembodiment seems
to get more training ground than the art of making good choices
regarding wise behavior. In the context of dualistic thought,
there's not much choice between good or bad, all or nothing. The
combined fact of conditioned disembodiment and fundamentalist
morality too often leaves us in a chronic, defended stance. This
has disastrous implications for our sensual selves.

Given the potent philosophers of the Renaissance and the scientific revolution, of the Enlightenment and the age of reason, we seem to have sacrificed the embodied, sensual realm in favor of the mind, the ideals, our conceptualizations. Our theories and explanations about the nature of reality have flourished while the body's direct sensual experience of it has slipped out of awareness. Without awareness of the body's response to each place and moment, our experience is little more than "a view from nowhere."[13] Without reading the internal signs, the subtle flutters and pulses in our bodies, there is no embodied sensation and little real recognition of being anywhere in particular. Every place feels more or less the same. The distinctions are lost and wherever we are becomes close to nowhere. We move on.

When awakened, perception is motivated, like a hunger of the body. And like lovers, our sensing and sensual bodies are fed on sound and scent, feasted by late afternoon light. Because we hunger for the eroticism such sensation affords our bodies, we are pleased to be called out of ourselves by the scent of a wild rose, the stunning yellow spike of goldenrod—and we answer back. We respond naturally. But within a landscape of man-made things, the toasters, travel trailers, "telephone wires, buildings, light bulbs, ballpoint pens, automobiles, street signs, plastic containers, newspapers, radios, and television screens" do not engage us. Rather, Abram says, "all begin to exhibit a common style, and so to lose some of their distinctiveness. . . . The mass-produced artifacts of civilization, from milk cartons to washing machines to computers, draw our senses into a dance that endlessly reiterates itself without variation."[14] We are soon dissatisfied by the lack of luster. Looking away, we search for that which shifts and changes, that which holds mystery and corrals our interests. Unfortunately, in a landscape of man-made things, that which changes is merely *new.* We

buy what is new and throw out the resources we have already ac-
cumulated. They are now nondistinct stuff.

But we do not see even this. We do not see the sensory body in-
nocently colluding in materialism arising from sensual dissatisfac-
tion. We do not fathom the disappointment in the body. We call
it depression and think that it is in the head. We read it as a sign
to see a therapist. We imagine our condition to be merely personal, perhaps biochemical: "Ah, to have it simply taken care of,"
we think, as we contemplate Prozac like a fast-food fix.

It is this sad fate of the body that James Hillman refers to in
*We've Had One Hundred Years of Psychotherapy and the World's Getting
Worse:* "The ugly makes us neurotic." He adds that "the special role
of the psychological citizen is the awakening and refining of aesthetic sensibility."[15] We have yet to develop a discriminating sensitivity in the body, a finely tuned visceral sense. We do not
readily distinguish between our responses to the synthetic, mass-
produced stuff that our corporate wizards insist we need and our
responses to that which has the form (and beauty) that engages us,
that naturally responds to the hunger of our sensory bodies, our
sensual selves. In our present condition, it all looks and feels the
same to us, too much the same to us, the same to us. Like a broken
record, we compose and live in a two-dimensional reality gone
dry. Numbed out to the max, we blindly wander through the mall,
gathering. We buy a blouse we never wear. What were we thinking? What were we seeing? Not much. And where the heck was
the body, anyway? Apparently nowhere.[16]

The roots of our disembodiment run parallel with our history of
rationalism. Plato's celebration of the mind over the body was
clearly an early influence on our present state. From Plato's perspective, mind was infinitely closer to the Ideals, to God, than

was any material form, much less a sensuous human body. From Plato we get "platonic," meaning we don't kiss. The mechanistic ideology that eventually arose from rationalism and the great technological advances of the Renaissance, in collusion with reductionism, rendered a mechanical view of the body. The body became predictable, stripped of its own intuition, its visceral recognition. It was now a thing to be treated by the developing medical profession, the new authority. In the process—with the burning of witches—we laid down the nettles, the slippery elm, and much of our wisdom in the way of herbs. We also relinquished much of the wisdom of our own bodies—the felt nuance, the gestures and intuitions running through the organs, the reach and pull of a big moon tide, the scent of werewolf residing in the filmy relationship between a full moon and viscera.

But I notice my heart quickens as I walk past a new clearing in the woods. It was a stand of pines, now sacrificed for someone's view. My breath shortens and my heart begins to ache, as if for a lost love. When I finally recognize what my body is doing, I wonder if my response to the killing of trees reveals the source of so many heart attacks, the heart pained and constricted. I wonder if the prevalence of heart attacks in industrialized countries is a measure of the numerous and daily destructive acts we perform, all of which require that we close down, numbing ourselves, numbing the heart, the very center of our pulsing bodies.

Words like *disembodiment* and *mindlessness* imply that we have vacated ourselves. Are we out of our minds? Is that what is reflected in our thoughtless consumption of the earth's resources? Are we so distracted by the stuff of the world—our minds filled by the imagery of the stuff to be owned and by the busyness of the stuff to do—that we are simply pulled out of our better judgment?

Ellen Langer, the Harvard psychologist who has popularized the notion of mindful learning, characterizes mindlessness as "an entrapment in old categories; by automatic behavior that precludes attending to new signals; and by action that operates from a single perspective."[17] But what are these old categories? How do we define automatic behavior, and what's the single perspective? If we allow Western civilization to define these terms, the categories tend to be black and white; automatic behavior is about individual interest, and the single perspective seems to be our own. On the surface, this may sound overly simplistic—my analysis itself cast in black-and-white terms. But among the possibilities for deeper discourse on Western varieties of mindlessness, projection fits the bill. Projection is definitely an either/or, black/white event. It is soaked in self-interest, and if the single perspective is our own, then, by definition, we are stumbling through a projected world. Consistent with Ellen Langer's carefully constructed definition, projection appears to be a classic form of mindlessness, and it's not simple.

From a Freudian perspective, projection is an ego-based or self-centered tendency. Projection takes the form of unconsciously casting ourselves—and particularly the parts of ourselves that we would rather not call our own—onto another. For Freud, "another" was always human. But projection is a nondiscriminating habit. We cast our shadows, our unconscious needs and assumptions, all over the place, as if throwing our minds into the world like a wet blanket, covering the reality of whatever we are looking at. In this way, we are out of our own minds, mindless, and yet somehow presuming that we understand what we're looking at. And in a sense, we do—we're looking at our very own assumptions, our unconscious needs, and our relatively comfortable, categorical take on the world.

The problem with projection, of course, is that there is little chance of seeing the truth or actuality of "the thing seen." For example, despite his protest, I tell my lover that he doesn't love me. Is this more accurately a statement about not loving myself? Am I casting my own self-doubt onto him? Is this a simple projection? In my desire to feel close to the natural world—to a tree, for example—do I end up telling the tree what the tree is telling me? Did I really hear the tree's message or do I simply supply it with what I wanted to hear? Projection may be a safe strategy for the ego—we get our own, albeit distorted, way—but projections are often inaccurate and sometimes just plain wrong. Furthermore, there's no chance of surprise or wonder when we view the world through a projective lens. We've already scripted the story.

Projection arises from dualism and is also heavily infused with doses of individualism and competition. It's win or lose, you or me; not both. And winning means that my reality is more real than yours. This kind of projection easily becomes a form of interpersonal domination, an attitude magnified by our history, including all the imperialist acts that our culture has propagated and rewarded. But ultimately, projection appears to be based on need. It is most infused with insecurity, as if we are trying to grasp the world in self-made, self-similar, and self-reinforcing terms. In an uncertain and competitive world, and with fundamental and unspoken questions about our place in it, this comforts us. We see the world the way we most want it to be. The lover even affirms my uncertainty. At least I know I'm right.

But the needs that secretly motivate our projective habits are deeper than needing to be right or wrong. They reflect a sense of alienation arising from what cultural historian Morris Berman calls "the basic fault,"[18] or what Chellis Glendinning calls the original trauma.[19] From Berman's perspective, the basic fault is the emptiness that pervades our subconscious lives and follows from the historical gap between our contemporary experience and the kind of natural experience that was typical throughout our long evolutionary history. The original trauma refers to the advent of domestication and to the associated historical events that nudged us further from engagement with the natural world. Both concepts ultimately refer to our deeply evolved hunger for engagement with the natural world, our bodies, hearts, and minds fully participating. They are referring to an unquenched thirst for an embodied sense of being at home in the world. Given an evolutionary history of perhaps 60 million years and a distinctly human history of at least 1 million years,[20] living in the midst of modern and mundane conditions leaves our bodies feeling as if they have been cut off at the roots, thirsty for satiation. Thirsty to be used.

With our senses cut off from a deep engagement with the colors and sounds of a dense and vibrant life-world, we become increasingly disembodied. Even the body knows that it is easier to forget its own capacity than it is to continually feel its ancient yearning. The depth of this yearning is yet another route to disembodiment. According to Berman, such disembodiment translates into projection. In some sense, we become mindless when we lose awareness of our bodies. Divorced from them, we do not know our whole selves. In our desire for full self-knowledge and in the imbalance and uncertainty that arise in the absence of an embodied sense of being at home in the world, we feel an amorphous need to know something. We provide labels and want facts reassuring ourselves that at least we know *something*. At least we know we are right. We feel more in control this way. But our readiness to label the world is, again, a shallow disguise for basic insecurity.

Our readiness to label the world alienates us from what is truly before us. In the context of projection, it is as if the alienating process, as Freud originally surmised, is consistent with the content. The content of projection is the shadow, the disowned parts of our selves, now externalized. In the process of actively disowning ourselves, we unconsciously thrust ourselves into the world. But the act of projecting is the externalization of what we most need to see in order to regain our wholeness. In the act of projection, what we cannot see about ourselves becomes revealed, mirrored for us. We see ourselves in the parts of the world that we disown, label, and do not allow their own existence. Consistent with the dynamics of projection, these parts are often labeled in egoic terms, meaning "for our use." A good part of the world is now labeled "resource." Resources have become something we both use and somehow know we are overusing. In this way, our projected shadows have colluded in the degradation of the world.

In a social system where the most competitive individual is the one who most succeeds, the insecurity at the root of projection must translate into a form of fear. And as we have seen, through the lens of the fear-based ego, the world is tinted with a brand of self-interest that strips the visible Other of its own substance and symbol—rendering a kind of blindness. This must be what the Australian Aborigines mean when they say that fear blinds us. In this case, fear and blindness certainly go hand in hand, as if joining forces.

The story of projection is the story of a runaway system. Labels and assumptions become amplified in the mind, in the field of our awareness. Take, for example, the adage that blonds have more fun. We project this, pouring our assumptions into the field of relations. Our view becomes so filled with our own imagery that we do not see what is before us, perhaps a sensitive and angry young blond woman. Instead, we confirm our bias, amplify the assumptions, and never see the world as it is. The resulting view is self-reinforcing and quickly becomes a "runaway system." In systemic terms, a runaway system is far from equilibrium and so has the greatest potential for either breakdown or breakthrough, and the outcome is unpredictable. This recognition makes me hopeful.

Projection is but one variation on the theme of Western mindlessness. "Our attention seems to be everywhere at once," says Chellis Glendinning. She is suggesting that with our attention thoughtlessly scattered around the face of the earth, we are, again, out of our minds. Intentionality, a considered presence of mind, is a word missing from the training ground we call education; imagination, a particular capacity of mind, is generally thought to be idle distraction; and mindfulness is something the Buddhists practice—we don't seem to have time. As an entire species, however, we seem to be at a critical juncture—will it be breakdown or breakthrough?

A Riff on Recovery

The world is not with us enough. O Taste and See. . . .

—*Denise Levertov*

A quick look at both our historical tradition and our contemporary environmental conditions makes it easy to see the sources and implications of our Western ways of seeing. It is also easy to see an absence of mind and body, and that we seem to be shredding the tapestry of creation with our object-oriented eyes. All of this suggests a desperate need for a new view. I look to the salt marsh before me. It is not a thing, as emphasized by "a marsh," a noun—meaning static, inert. It is not even a location, as on a two-dimensional map. It is many things happening, an event, an orchestration, an entire symphony. It pulls me from a mind made busy by habits and codified desires and guides me into a subtle slow rhythm, into the daily tidal pulse caused by great gravitational bodies in orbit. The incoming tide gently heaves and swells, rolling up and into the marsh—a fingering of land and sea, followed by another low wave, and another. Water moves across and through the skin of the earth, the waves lapping up against and passing through marsh grass and newly blooming sea lavender. The entire marsh sways with the sloshing of ocean, and I feel the earth turn.

Almost invisibly, a warm fog moves across the sea to marsh. The sheen of water and thick marsh grass are soon obscured by gathering volumes of soft haze. I am reminded of what we often do not see: the pale crescent moon in midafternoon, tide pools at high tide, hearts pumping just below the surface. To limit our faith to the visible facts is to call our world a done deal—no mysteries left, no revelations to be discovered. Our recovery must weave the unseen realm of psyche into the material moment, restoring our faith in the embodied irrational, the blush and subtlety of maybe, the meaning of mythic. The mystery of mind must be called into the light of day, unabashedly rejoined with the visible world—which without our gift of imagination is too easily reduced to stuff, nonrelational and ultimately uninteresting. Recovery must begin with noticing what's truly there, with looking into and between, with revealing what has been left unseen, and with an offering of wonder and imagination.

So the crucial question remains: What do we bring to the marriage of inner and outer realms? Can we awaken, or is it too painful to do so, to be so? Do we choose the superficial comfort of closing down our perceptual channels, defending ourselves but simultaneously missing the juice and vibrancy of the world? Or do we look right into what is truly before us, the whole scene stretching between the sweet sublime and nasty, god-awful reality?

For now, just look, and look again. This, then, is to respect. We know the word, but when we choose to truly *live* it—to look again with sincerity and depth—the heavy, dull sensation we call "numb" slips away across the landscape like a cloud shadow in wind. We see the vast red rock emptiness left behind, windblown and raw, bared to bone. Bared to the bone, we too become raw and vulnerable. We often regress, slipping away from the world, hiding and peering from the very crack that initiates recovery. Soon depression leaks through, signaling blue, blue sky and weather. Weathering the blues, we look to sky, wide and now blazing with stars and dark, deep, the vast unknown. Is it wonder or terror we see there? Are we ultimately at home in the world? Or are we still adolescent and angry? Do we hiss through the throat of repression, the ancient losses still unspoken? I hiss, like Snake. I see myself hissing in a fine meeting of colleagues. There is no visible reason for this, but it is a sure sign, a signal that I am uncoiling, shedding, breaking the crusty shell of defense. I crack open—we crack it open—cracked open by loss. The poignant loss of a beloved world, drowning in toxins. Of a salty marsh, of the fish swimming there, drowning in toxins. We crack open, dense with true desire.

Chapter 3

Mindful Eyes

Seeing as If the World Matters

Wisdom! Attend!

—*Huston Smith*

Attention, Our Offering

The energy it takes to develop attention, which seems exhaust-
ing at first, returns to us as radiance and reaches into every level
of life. Nothing we do is lost; everything we do has greater
aliveness, greater freedom.

—*Richard Moss*

I am riding a mountain bike up a steep hill in the Bradshaw Mountains with my friend Julie. We are both pumping hard, but it is a
brilliant day and we feel brilliant, too. The top of the hill is chunky
with granite rubble, where a fallen ponderosa pine has suddenly altered the trail. My attention is on the right-angle jag in the trail
and all that nasty rubble. By the time I get there, I ride directly into
the mess. I ask Julie, who has slipped through the chunky stuff and
is still pumping, "What's attention?" She throws a quick glance over
her shoulder and says, "It's alignment." I know what she means. It's
the fundamental lesson in mountain biking: Look where you want
to go. In other words, align your looking and thinking, or even better, your gaze and intention. The body naturally follows.

We call this kind of alignment "*paying* attention." Our language
suggests that attention costs us something. Perhaps we liken it to
money because it takes time. Time is money, right? Attention, so it
follows, is something we pay for. But with a consciousness conditioned by a market economy, it would appear to be grossly ineffi-
cient to pay for attention; there is nothing tangible to be had in
return. Perhaps this is why we don't seem to do much of it. On the
other hand, attention is the source of what environmental writer

Robert Michael Pyle calls "experiential wealth," and certainly the experience of "paying" attention is the flip side of psychic numbing. Within psychology the most common definition of attention is "the enhancement of selected sensory information." Focused attention, we know, translates into a richness of color, a depth of sensory experience, and the difference between seeing and not-seeing.

Attention has hundreds of definitions. It has also been periodically shunned by research psychology for its lack of precise definition; yet it is easily recognized by anyone who cares about anything. Ordinary attention (the kind that everyone has an opinion about) moves from object to object, focusing, letting go, moving on. It is prone to practicality, with a focus on whatever we need to see in order to get the job done. But our attention is both easily seduced by passivity and fickle. It's there and not there, skipping through consciousness with unconscious abandon, with the unruly intensity of a hound dog. Ordinary, untrained attention, says philosopher and religious scholar Philip Novak, "comes and goes without our consent; it is not something we *do*, but something that *happens* to us."[1] Nonetheless, it serves us well. Even unskilled attention brings much of the world into the focus of our awareness.

"Everyone knows what attention is," said William James. He described attention as the mind "taking possession" and making a single object or thought (among many possibilities) the most clear and vivid. Simone Weil described attention with a distinctly different flavor: "Attention consists of suspending our thought, leaving it detached, empty, and ready to be penetrated by the object. . . . Above all, our thought should be empty, waiting, not seeking anything." In *A Blue Fire*, James Hillman writes: "Attention means attending to, tending, a certain tender care of, as well as waiting, pausing, listening. It takes a span of time and a tension of patience."[2] Like the notion of "just seeing,"[3] Hillman advises us to keep our mouths shut, to refrain from being know-it-alls.

Attention must be cultivated, gently courted, as if we are courting not only the world but ourselves too. We ask, are kind, we give of ourselves. In return we are granted especially rich glimpses of the world. The "ten thousand things" that make up the Buddhist view of the world become both more distinct and closer, somehow more intimate, as if making themselves available to us. Initiated by our intention, attention offers a kind of coming together with the world. Over time, our gratitude rises up like a wave surfacing from still water, meeting the world in a purely present moment. Contrary to the act of *paying* attention, there is grace in the offering of our attention.

My thesaurus lists the synonyms of attention as "focus, key, center, crux, feature, highlight, spotlight, awareness, recognition, care, carefulness, cognizance, consideration, heed, intimacy, knowledge, knowing, mark, note, notice, observance, observation, perception, regard, remark, sense." This wide-angle list of synonyms suggests that a rough taxonomy of attention would at least include "the experience of" attention, the phenomenology of it as one major class and the "theory" of attention as another. The theory of attention is concerned with what it is and how it works. The phenomenology of attention is a different matter. Even the unskilled observer experiences the sharpness of its focus, the cost of its absence, and the way the world is created anew with a single attendant glance. Attention, we know from our experience, drifts, lasers, and slips away again. It can be focused on the body, the contents of mind, the stuff of the world; it can inhabit the toes, reveal the heart chakra, amplify loneliness, or penetrate the neighbor's backyard.

In Sanskrit, the concept of attention is expressed as *dhyāna*, "the long, pure look." Dhyāna is a concept born of direct experience and includes both spontaneously focused attention and the sustained practice of attention through meditation. These two forms of dhyāna differ greatly. One is transient and involuntary and rebels against focusing on any one thing. It is a pure and wild form of presence in the world. The other requires much discipline, is highly voluntary, and, ideally, is sustained. But in both cases, dhyāna recognizes that one loses one's sense of self in the moment of attending. As writer Padma Hejmadi says,

> At some point you are seeing so intensely that you become what you see, you merge into the drop of water until the "you" disappears. The how and whys and wherefores disappear too. Yet when you emerge, you are somehow replenished.[4]

"True attention," says Flora Curtois, author of *An Experience of Enlightenment*, "is rare and sacrificial. It demands that we throw away everything we have been or hope to be, to face each moment naked of identity, open to whatever comes."[5] Perhaps it is this quality that marks attention for what it is. Perhaps it is the fact of the world showing up vividly, with clarity and depth, as we give ourselves away, as self is relinquished, that most characterizes what we call attention.

In general, attention for the contemplative is an active engagement, an "effortless effort" to observe, to sustain the observation, and eventually to suspend attention itself and simply be "all there." In *samatha*, a Buddhist practice, the capacity to sustain one's observation is tuned, not mastered, by aligning one's intention with objects of meditation, a *koan*, an image, one's breath, the contents of consciousness. For the beginner, attention is taught as if one were developing a skill, although ultimately the true practice of attention is designed to help us unlearn mindless habits. The goal of samatha is nonreactive acceptance of whatever presents itself in the busyness or stillness of one's mind, the eventual cessation of thinking, and a full awareness of the present. The experience is one of spaciousness. In a moment of "true attention," self-consciousness fades from view, as if too small for such spacious attention. Even early in the meditative practice, the observation of the endless train of passing thoughts and feelings—the contents of mind that superficially satisfy our needs for identification—reveals the self as relentlessly impermanent.

Given our Western sensibility, the idea of the self as "relentlessly impermanent" is problematic: We don't dare lose ourselves. And considering that our linguistics tempt us into thinking that attention costs us, it is not surprising that Western inquiry has comfortably focused on the quantifiable nature of the objects (albeit reduced to stimuli) that capture our attention or on the mechanisms embedded in neural circuitry. Attention through a Western lens is theoretical, complex, and much removed from meditation or mindfulness. It is not a phenomenological inquiry, nor is it studied in the interest of developing one's capacity for attending. The scientific study of attention, however, is useful in the interest of altering perception.

From the perspective of visual science, attention is thought to be both an internal process marked by "capacity" and a focus or beam that shifts across space. The language of attention as it is used in visual science reveals a primary interest in how it works. In spatial terms, attention has been metaphorically described as a spotlight, a hemifield phenomenon, and a task-dependent zoom lens. It is "deployed" voluntarily and "captured" involuntarily, and in both cases it mysteriously enhances signals selected from the vast realms of external and internal landscapes. In relation to its selective capacity, the metaphors commonly used for attention are filter, bottleneck, and gate. The traditional way to explain its function has been "the cocktail party effect," reminiscent, I suppose, of the 1950s when attention became a spicy research topic and when such vocabulary was, I presume, what the research crowd could easily identify with. The cocktail party effect describes our ability to suddenly hear our name "pop out" in the midst of noise, the buzz of cocktail conversation. This expression was also a reference to a good deal of evidence supporting the notion that our familiarity with a certain stimulus was the greatest predictor of what would receive our attention.

Beginning in the late 1970s, many studies investigated the relative "costs" and "benefits" associated with the selective capacity of attention. The overall conclusion was that the familiar or expected signals were seen *much* faster, and with *much* more accuracy, than anything unexpected or unfamiliar.[6] Although quantifying the value of attention was big news, it's not the whole story. For those currently doing and reviewing large amounts of research on the topic, attention is "the process that brings a stimulus into consciousness."[7] In other words, it is what allows us to see in the first place. Without it, we are "inattentionally blind."[8] It's a substantial phenomenon, the difference between seeing and not seeing.

Within neurophysiological research, attention is classified as both "endogenous," internally generated, and "exogenous," that which arises by virtue of the dynamic, demanding world of innumerable things buzzing, ticking, changing, and jumping into view. Exogenous attention is what I imagine the extrovert to be especially good at. Exogenous attention is what happens when something out there solicits and seizes our senses. Our attention is suddenly and involuntarily riveted on, or drawn to, whatever has the power to capture it. From an evolutionary perspective, this kind of attention focuses our senses on threats or dinner with a speed driven by survival. It is hearing the snap of a branch and spinning around, catching the flick of tail, locating a lion in split seconds. Fully awakened, with the attentional gate both flung open and ready to snap closed, it is fast and flexible. It grants us the flash of red on black wing, that hint of smile, this shiny stone in wet sand, that sexy riff of birdsong—all suddenly emblazoned on the brain. It surprises us and demands that we look and listen.

For instance, I am in the Grand Canyon, alone and stretched open wide, like the Tonto Platform I am traversing, a thousand feet above the Colorado River. The sky is a nasty gray, full of wind and enormous thunderclouds gathering quickly like bats out of hell. I am walking fast to beat the storm—until I am stopped short, my eyes captured by and riveted on a particular something over my left shoulder, a spot of ground. It looks like all the rest of the Tonto blackbush-covered landscape. I see nothing that warrants such attention. But so demanding is the call to look that I hunch down, circle around, and slink up to whatever-it-is quietly, like an animal carefully stalking prey. As I draw near, I see a circle of stones, maybe a dozen feet in diameter, just barely visible in the brush, now grown over and scattered by many years and much weather. All I see, really, is mystery. The only thing I know for sure is that I have just experienced the power of my attentional antennae. Later, I am told that such circles are believed to mark the picnic grounds of traveling Anasazi.

Endogenous attention, on the other hand, refers to the way in which we are primed to see whatever resonates with our own cortical landscape—what we expect, assume, or desire. In functional terms, this capacity has been called our "perceptual set," but we might think of it as a kind of loyalty to our personal history, to the tracks laid down in an otherwise chaotic mess of neurons. Endogenous attention arises from an honorable attempt on the part of the brain to make sense of, to organize the ample, enormous world of signals. It is thought to be the outcome of both neural networks laid in the cortex and the fact that there are many neural fibers coming down from the visual cortex and influencing incoming signals from the retina. Incoming signals are mixed and matched with top-down influences at the lateral geniculate nucleus (LGN), in the center of the brain, before flashing on to the visual cortex and other brain sites. We might think of the LGN as the place where a vast catalog of neural baggage and business meets the visible world, in the form of biochemical signals.

What we perceive depends on where we place our focus of attention. The conscious or unconscious placement of cognitive energy on internal desires, needs, and priorities acts as a filter by selecting information from the visual field that resonates, that "patterns up" (presumably in the LGN) with the already humming neural activity. For instance, when I'm hungry (and humming along inside with the memory of muffins, eggs, beans, and blackberry pie), restaurant signs "pop out," catch my attention, and call me over. Along the same lines of self-interest, endogenous (inner) attention also filters and focuses on what we need to see. Oh, yes, poison ivy, I suddenly think as I walk along a path that has always been lined with the nasty stuff. In the next split second, great patches of shiny, red-tinged leaves blazon into view. Although I hadn't seen it until this moment, poison ivy glistens everywhere. But unfortunately, endogenous attention works in reverse as well. My first two weeks of tracking baboons in Tanzania were most remarkable for what I couldn't see. Even the big animals—giraffe, eland, and zebra—defined my view. Since I had no firsthand familiarity with African wildlife—and no doubt little, if any, neural networking to support my vision—they easily escaped my eye.

Despite the different qualities of attention, the fact that it selects, enhances, and essentially creates our reality cannot be underemphasized. And there's the crux. By filtering the visual world through neural substrates representing our previous experiences, endogenous attention builds and perpetuates our subjective reality. This is both problematic and useful. By selecting what information is perceived with reference to our expectations, immediate needs, and familiarity, attention affirms and perpetuates habitual assumptions, stereotypes, and, quite possibly, a worn-out worldview. On the other hand, attention, in concert with the power of imagination, has the capacity to create new visions of reality.

Attention, whether focused internally or externally, is an exceptionally dynamic process—fluid, flexible, and generally assumed to be automatic. And because some degree of attentional focusing is automatic, we generally take our ability to attend for granted. But there's plenty of psychological evidence to show that the ability to attend is a learned skill, and it's certainly apparent in learning to ride a mountain bike on gnarly trails. Within the great wisdom traditions, the cultivation of attention is central to devotional practice. This is most apparent within the major contemplative schools of Buddhism, where the cultivation of mindfulness requires (or is identical to) the practice of attention. Within the Theravada tradition of Buddhism there are essentially two forms of practice, *samatha* and *vipassana*. Samatha is the practice of "one-pointed" attention. Its goal is the understanding of certain Buddhist doctrines through direct experience, through the observation of "objects of meditation" and the processes of the mind. The goal of vipassana is the development of insight. According to tradition, insight may be developed only after one has first cultivated a highly concentrated form of attending through samatha practice. The Tibetan practice of visualization also follows samatha, requiring the mind to create and maintain elaborate images of mandalas and deities as a way to foster sustained attention. The two practices central to the Zen tradition, *koan* and *zazen*, similarly require the practice of sustained, wakeful attention. The koan provides a mental focus, a conundrum that cannot be solved by the rational mind, and so pushes the practitioner past thinking. Zazen, or sitting meditation, requires the practitioner to simply attend to what is in the present moment. Nothing added. Just here, now. Attend.

The implication inherent within these long traditions of devotional practice is that attention is something to be developed, cultivated, learned. And the learning curve is steep. This became exceedingly obvious to me while doing research on attention. I arrived as a visiting researcher at the Smith Kettlewell Laboratory in San Francisco, very excited by the invitation and naively blind, "inattentionally blind." For the first week or so, I saw few of the many "targets" that were presented on a computer screen—despite the fact that my attention was cued and the targets were high-contrast stimuli of long duration. In the midst of both men and science, I felt like the classic dumb blond—only in this case, it was blond and blind. But after a couple of weeks and hundreds of trials, I became very good at seeing 32-millisecond presentations of gray stimuli on gray backgrounds. Manfred, who had been running himself through experimental trials for over a year, saw everything—every single split-second, low-contrast target. Ken, who masterminded the experiments, interpreted data and never sat as a subject, never saw anything. His conclusion was that Manfred and I had screwed up the program that ran the experiment. It became a good joke.

But this sort of anecdote is rarely, if ever, considered to be "proof" within visual science. The development of explanatory theory, through statistical calculations of data, is what drives the scientific community. Besides, we were in the business of measuring the "costs" and "benefits" of the normal observer. Our research traditions do not readily lend themselves to studies in human potential. Subjects who perform outside the norm are statistical nightmares and are inevitably discarded from the subject pool and database. They are referred to as "outliers" and "practiced." Within visual science, "practiced" subjects are frequently replaced. But the very fact of subject turnover is evidence, in my mind, that we *learn* to attend. If that is true, then we may learn to truly see.

The word "attention" is derived from the Latin *attendere;* "tendere" means "to stretch." Flexible, we move out into the world.

Never the Same

> His seeing was never the same again.
> The connections between things had changed.
> —*Padma Hejmadi*

The visual system is made up of neurons and neurotransmitters, well-worn neural paths, associations and expectations. It is both hard-wired[8] and soft, pliable, and plastic. In my materialist-gone-mystic mind, it is this plasticity that most fascinates me. Most fundamentally, plasticity refers to changes in cortical structure as a function of experience. Structural changes are alterations in the degree of connection between neurons. More highly connected neurons are referred to as "facilitated." The connections, or synapses, between facilitated neurons require relatively little incoming energy to fire and thus pass a signal further into a neural network. Because each neuron has thousands of connections, the firing of a single synapse ripples exponentially into the brain, choosing to pass through the most open gates as it flows. Neural networks are the latticed patterns of activity, the "resonating neural ensembles,"[9] and are presumably the neurological stuff of recognition. Another way of saying this is that the recognition of a visual object or attribute results from the emergence of a coherent pattern of activity among resonating neuronal ensembles.[10] In the language of psychology, neural networks thus constitute our schemata or prototypes, which automatically determine the ready categorization of visual input—and in the case of the relatively unfamiliar, our ability to perceive at all. Taken one more step, changes in the connectivity between neurons alter the flow and emergent patterns of activity, representing categorically different ways of seeing. Plasticity thus refers to the rearrangement of synaptic connections, to the flexibility in the system, and to seeing the world differently.

Despite a deemphasis on changes in perception within the community of visual psychologists, there is a fascination with visual system modification among neurophysiologists and cognitive scientists. Following David Hubel and Torsten Wiesel's identification of the structure of the visual cortex, a good deal of the research in visual neurophysiology was focused on identifying the developmental phase during which significant changes in visual system structure and capacity occur. Those who investigated what came to be called the "critical sensitivity period" assumed that cortical changes were age dependent and kittens and young monkeys became the research subjects of interest. But there were also adult monkeys and cats who showed anatomical, physiological, and behavioral changes as a function of experience. In fact, there were quite a few of them, although their changes were less predictable, more subtle, and certainly less sensational. As an undergraduate student, I asked about these cases. One rather stern professor peered at me as if I were up to something dangerously good, said nothing, but submitted the paper in which I had quite seriously posed the question to a university awards committee. Although I did not win any prizes for questioning the canon, this professor did comment in a letter of recommendation that I had "a nose for the important questions." She had been studying the development of infant vision for many years and I suspect she knew that developmental processes and learning were methodologically mixed together, or, in scientific jargon, confounded.

Later, when I asked my favorite scientists at Brown University, the men doing what I thought was the best research on visual system plasticity, they too smiled and said little. But in private, a couple of them did admit to me, "Yeah, they're there." They meant the experimental results showing changes in visual system structure in adult cats and monkeys. Whenever I brought up the possibility of adult plasticity in visual neurophysiology seminars, however, the same researchers nodded and smiled as if to say, "Don't ask." I remember having a distinct feeling that, yes, I was on to something good.

In addition to these clues from professors and researchers, I had experienced considerable changes in my own visual capacity several times and had seen numerous others go through radical changes while teaching vision improvement. I also knew that there was a vast literature on the development of visual literacy and a sizable one on practice effects in perceptual psychology experiments. To me, the development of myopia seemed to be a similar phenomenon, except in reverse.

The upshot of all this is the fact that vision changes. From the perspective of cognitive psychology, perceptual learning is no more than the shifting of resonances. A gradual change of connectivity between neurons is modeled on the Hebbian learning rule. Basically, the rule states that the strength of connection between neurons changes with the simultaneous firing of both neurons, when one neuron passes a signal to the next. This connection is strengthened when more signals pass through and atrophies in response to a decreased level of activity.[11] If the connection is strong or facilitated, relatively low level signals can trigger the neural network, the system, to fire. In terms of experience, this presumably means being able to see (or at least identify) with a mere glance what we have already seen many times before.

Applying the Hebbian learning rule to neurophysiological findings, researchers at Brown University produced a model of cortical plasticity, of restructuring within the visual system, in response to incoming signals.[12] The critical factor in their model was not age or developmental phase, but rather the value of a threshold that determined the level of activity required for modification to occur. The value of the modification threshold reflected the degree of connectivity between neurons and shifted in response to the activity of noradrenergic and cholinergic pathways.[13] To make a long biochemical story short, these pathways are thought to be activated when attentional mechanisms are awakened. The model, based on many years of neurophysiological findings, implies that plasticity is not simply a matter of developmental phase. More essentially, it's the act of attending (the amount of "arousal," in traditional neurophysiological terms) that makes the critical difference. It is being as excited as a child, and mindful as an adult that makes all the difference.

This is key: Our attentional focus, both internally and externally, influences and creates subjective reality by facilitating the perception of some objects, relations, and events to the exclusion of others. This happens instantaneously when a previously formed neural network of associations and the consequent selective control at the LGN are activated. And it happens over time when specific neural pathways are altered by being repeatedly activated and thus facilitated. We might call this the hermeneutics of perception, or true karmic causality with reference to perception.[14] Edward Hall, an anthropologist-philosopher, put this more simply: "Man learns while he sees and what he learns influences what he sees."[15]

In behavioral terms, this means that our way of looking at the world, our habitual focus of attention, determines the structure of our neuronal ensembles. This, in turn, determines what we will see, whatever resonates in the mix between inner and outer landscapes. Because resonance is especially easy to come by if the neural net is well traveled, almost any glance sets the stage for seeing more of the same. Our habitual way of looking at the world is iterative—in effect, it accumulates our worldview. This is how our perceptual habits, conscious or unconscious, become our reality. Another way to say this is that we tend to see the world as we believe it to be, unless we consciously choose to attend to what we haven't yet seen, to notice the unexpected elements of the world.

Ultimately, attention is the true power of vision. It is the capacity to join inner and outer landscapes, to bring the contents of the mind and the things of the world together. It is alignment. The focus of our attention may be placed within inner and outer landscapes to various degrees of depth and levels of scale, and with many degrees of intensity. In the process, we choose, enter into, retreat, open, lose ourselves, emanate and radiate. Attention weaves together mind and matter and so inevitably determines the quality of our interaction with what the Buddhists call the "ten thousand things." We don't perceive anything without it, yet unwittingly we often let it languish, diminish, go unconscious. And, within contemporary Western culture, there is little or no training in the art of it. But we all know from our own experience that an offering of one's attention requires care, something like a fullness of presence. Maybe this is why we are so unskilled and often don't see or hear much of the magic that is all around us. It is hard to generate our full presence in a world gone awry. And if our careful attention is the ultimate form of being (as is thought by Buddhist practitioners), then perhaps this is why the Western mind struggles so desperately with identity, with the question of who we really are.

Mindfulness Turned Inside Out

Why is the mind incapable of deciding its own subject matter?
—*Jeanette Winterson*

Bhante Henepola Gunaratana, a Buddhist monk and author from Sri Lanka, says in the article "Mindfulness and Concentration," "Mindfulness picks the objects of attention and notices when attention has gone astray."[16] He adds that it is "a delicate function leading to refined sensibilities." From my Western perspective, mindfulness *is* the skillful use of attention. And given that there are elaborate traditions of mindfulness practice, such as Theravada Buddhism's samatha and vipassana traditions, I presume that mindfulness, like attention, is a learned skill.

I am told that lesson number one in any mindfulness practice is an awareness of our impermanence. Our sensations, desires, needs, and fantasies all pass continually before the mind's eye. We witness ourselves as no more than a continual and impermanent flow of experience. Only our attention can embellish a passing moment with the glitter of light, the velvet red of rose. Only attention can give depth and density to impermanence. Mindfulness practice next reveals a kind of spaciousness within each experience-saturated moment. Interdependent reality then becomes apparent in the blending of our own impermanence and the spacious and full flow of experience, of influence. We see our own passing identification entwined with the present moment, the thought spreading into sunlight, the scent of lilac breezing by, the shudder of a butterfly alighting. It becomes apparent that in a moment of deepened awareness, *we are* breezed into and shuddered by sight. In the moment of seeing and sniffing, we are entered. Do we shift in response? Do we *become*? Do I shift and change as light and birdsong enter me? Is this the ultimate goal of mindfulness practice?

In moments imbued with such presence we feel a subtle shift in our senses, and our hearts soon follow. These precious moments are wide-eyed with the welcome of the Other. This perceptual state is parinishpanna, the Mahayana Buddhists' experience of "ultimate reality," of perceiving the world without artificial dualities. There are no perceived categories in this realm, no divisions. We are the ever-present seer perceiving what is "always already," as Ken Wilber would say. With ever-present awareness we witness what is "always already" there: "Somehow," Wilber writes, "no matter what your state, you are immersed fully in everything you need for perfect enlightenment."[17]

In *Island*, Aldous Huxley's famous utopian novel, ubiquitous and unseen mynah birds continuously provide a mantra for mindfulness. They say, "Here and now, boys. Here and now." Practiced mindfulness is the capacity of the mind to be utterly present in one's immediate, everyday experience. The techniques of mindfulness practice are intended to "lead the mind back from its theories and preoccupations, back from the abstract attitude, to the situation of one's experience itself."[18] It is the development of "ever-present awareness." It is a simple concept that resists being simplified. It demands practice.

In the case of samatha and vipassana, the practice of mindfulness turns an intensified focus of attention toward the internal landscape. Buddhist practitioners may become aware of the contents of mind with such depth that what they call "the aggregates" or the "parsing of experience"[19] are observed. Cognitive neuroscientists hypothesize that the Buddhist experiences of "the aggregates" and the "parsing of experience" are direct perceptions of waves of activity passing through the brain as synchronized rhythms or resonant ensembles, rising and falling together in millisecond swells.[20] Visual mindfulness is turning this practiced and finely tuned awareness inside out. It is directing our attention to the visual field, outside of ourselves.

For most of us most of the time, ordinary attention acts as if it has its own mind, albeit a culturally conditioned one. It "comes and goes without consent." Even the most unreflective person must recognize that there is a kind of fickleness in our everyday attention. In some instances, we refer to this jumping around of attention as attention deficit disorder (ADD)—or more recently, attention deficit hyperactivity disorder (ADHD). At times, it seems as if everybody has it.[21] This fickle nature of attention is called "monkey mind" by those who practice Theravada meditation. But monkey mind, we know from experience, roams across both internal and external landscapes. The capricious quality of mindlessly casting our "spotlights" over the things of the world must certainly be an externalized form of monkey mind. Turning our attention "inside out" begins with simply noticing monkey mind at work outside of ourselves. Can we quiet the visual chatter, the relentless looking at everything and seeing nothing?

When we practice visual mindfulness we attend outwardly with the same kind of discipline that is fundamental to samatha and vipassana practice. These are concentrated forms of attention in which the practitioner seems to perceive synchronized rhythms passing through the brain, the "parsing of experience." We bring the devotional practice that samatha practitioners bring to the contents of mind, but to the color and curl of waves, the histrionic banter of gulls. We practice turning our attention to the ecological moment, turning our minds inside out, cultivating the capacity to focus and sustain our fluid energetic flux. We look, and look again, calling ourselves to the moment of the world, noticing what is.

But there is also a hidden dilemma in becoming mindful of the world as it is. Much of our world is now degraded. The fundamental lesson in mindfulness practice is the recognition of interdependent reality. In becoming mindful of the world, we also experience the ways in which the breakdown of the juicy, vibrant, green, and windy world translates through us. We also break down. In our despair, we witness our emotional bond, deep and rising now to the surface. We remember the fullness of our hearts, our capacity to care. "Attention *is* care," says master wildlife tracker and photographer Paul Rezendes. If attention is both our care and the true power of vision, if it has the capacity to co-create a world, then what do we attend to in an era most marked by ecological degradation?

I can offer a hint, a starting place. Here in the West, we are most clumsy when it comes to relationships. We have many cultural myths associated with radical individualism, and we navigate quite easily through the trials of getting more for our individualized selves, but we often slip up in the realm of relating. In the United States, divorce is at yet another all-time high, and we suffer child abuse and unprecedented violence against women and minorities. With even a sideways glance, it is apparent that we do not care much about our human relationships, and in general, we seem to have forgotten our relationships with the natural world.

Skillful Means, Receptive Means

It is possible, in deep space, to sail on a solar wind. The secret
of seeing is to sail on a solar wind. Hone and spread your spirit
till you yourself are a sail, whetted, translucent, broadside to the
merest puff.

—*Annie Dillard*

The Bodhisattvas are trained to use skillful means. They are Buddhist warriors on the path of enlightenment for all sentient beings.

They work in the name of compassion and with hard-won wisdom. Being on such a path, and having gone through rigorous training, they are sanctioned to "bend the rules." To be skillful, then, is to act wisely and with compassion and always in relation to what's happening in the moment. In other words, *skillful means* take into account what Buddhism calls dependent co-arising. Dependent co-arising is the fact of mutual causality within the world, the recognition that the ten thousand things arise as a function of relationship. It's the fundamental understanding of the Buddha, his central principle. Skillful means cannot be a simple list of rules—as if a single form of justice might guide ethical behavior independent of the relationships within which reality arises. Rather, skillful means are contextual.

Our era is shaped by "numb and not-noticing," projection as a common and unconscious habit, and ecological degradation. Within this context, skillful means must be receptive means. Receptivity is a synonym for sensitivity and noticing, the opposite of projection, and the ground of any real relationship. If this dawning ecological era is asking anything of us, it is that we notice the relationships that constitute the world—those that are degrading and those that are life enhancing. But there's not much chance of a significant relationship or of being ecologically—that is, relationally—aware when we're not receptive.

We don't talk much about receptivity. It does not appear readily in our vocabulary or in the all-American profile. It never shows up in psychology textbooks, despite the fact that it is fundamentally psychological in nature. Receptivity is rightfully linked to the feminine,[22] but in our culture we suppose that anything feminine is weaker than anything masculine. For this reason, receptivity is usually seen as a relatively negative trait, assumed to betray weakness or a lack of choice. Considering our cultural conditioning, however, receptivity must be a choice, an intention. It goes without saying that the prevailing pressures to compete, to be the rugged individual or the well-defended "good girl," do not foster receptivity. In this cultural context, we must make a conscious choice to open ourselves. Otherwise our reception suffers, as if our exquisite sensory receptors were clogged.

In our out-of-control (and controlling) Western society, it's obvious that skillful means must be contextual, relational, and thus borne of receptivity. At times, however, the context we find ourselves in suggests that being receptive is not wise. Skill becomes a matter of shifting degrees of receptivity, of reading the signs— with an eagle eye, with strong eye, with honed reception—and discerning when and where and with whom one can wisely open. When I go to the woods alone, I begin reading tracks the minute I leave the pavement and never let up until I've returned. Then I know who's around and what they're up to. The signs clarify how wide open I get to be. Fresh, man-sized boot tracks keep me mindful, wary, open-eyed, and vigilant. I assume an energetic stance, a particular form of receptivity aligned with a particular intensity of focus. But if I walk down a sandy wash all afternoon seeing only the tracks and traces of mice and raccoon, javelina and deer, I let the doors of my heart bang on the hinges, my body wide open in a gesture of welcome. I see the world through love eyes. Light and leaves and birdsong tug at me, at the organic core of me. I begin to resonate deeply, feeling myself rise to the occasion and become graceful, at home in the world and in the swing of things.

Ultimately, skillful means become a matter of grace. Are we able to open with ease when the occasion arises? Are we able to fine-tune the quality of our presence, shifting the form and degree of our receptivity? Are we ingesting and absorbing, retreating, or protecting ourselves with intention, flexibility, and grace? These skills are not difficult. The difficulty is that they are subtle psychological distinctions and that we live in a culture that too often allows us to be too unconscious, our psychological awareness underdeveloped and unrefined. We seldom talk about the nature of receptivity. Our conversations are most often limited to black-and-white, good or bad boundaries. And even though our senses long to be called out, we often don't know how best to behave.

Receptivity is a choice. My friend Antoine says, "You must first open the palm to receive." Receptivity is dreamy-eyed Hathor nursing Horus; she is in an altered state, giving and receiving in one gesture. It is Aphrodite, so given over to love, and the sacred prostitutes of ancient Greece, receiving and translating, taking the war out of men returning from battle. It is how I feel when I intentionally gather up the long rays of afternoon light in a wide sandy wash and, after several hours, feel the swirl of warm light in my solar plexus, the golden third chakra. As an open vessel, I am filled with light, en-lightened and made fresh.

The dictionary defines receptivity as either "able or inclined to receive ideas" or "fit to receive and transmit stimuli." Needless to say, this tells us nothing about receptivity in relation to the body or soul, or how it informs our pragmatic sensibilities. This definition, so inspired by our cultural blind spots, doesn't even give us a hint of how to do it. Annie Dillard says, "Center down." I say, "Slow down." Go to a place of beauty. I say, "Breathe," let your muscles do nothing and then let that sensation seep into your bones. Imagine your pores opening, your skin permeable, a thin veil between your heart and the world outside. I say, "Breathe." Soften your eyes, let your attention be wide. Listen, and then soak in color, and feel texture with your eyes. Give your attention to the golden glisten of pine needles in sunlight. Let your care roam over a landscape, flexible, both focused and soft. Let your practice be this simple and do it often, when you can.

Like attention, receptivity is an iterative process arising from an organic system. According to systems theory, living systems are self-organizing, organizationally closed, and energetically open. With a little conceptual stretch, *energetically open* means primed to receive vibrations. This is our natural, organic state. So imagine this: You peer around the culturally conditioned lens and step out of a culturally induced trance. You intentionally open your senses, opening to the vibrations, the color, the scent and feel of *beauty*. If resonance is the felt or embodied sensation of the vibrations associated with beauty, then, by definition, it feels good. You'll want more of it. With full respect to the Dineh,[23] you are now on the path of beauty, the Beauty Way. Because receptivity is iterative by nature, it easily accumulates itself, becoming easily strengthened with positive feedback, with intention.

In the context of the omnipotent "numb and not-noticing" perceptual stance of our times, I know this approach is skillful. When we mix our attention with good reception we truly see, and what we see is clear, deep, and dense. The signals soak into our neural circuitry, guiding our next glance, and the next. Soon beauty appears almost everywhere; the world turns on. With the body tuned in, we fully resonate. It feels good, the sensations become increasingly differentiated from the look and feel of the not so beautiful, the plastic crap and crud. We make no mistake about the path we are on. We're walking the path of beauty and there's no turning back.

But again, the world is not always beautiful. And so here's the skill: In the face of the clearly not so beautiful, we do not look

away or unconsciously close in a spasm of denial. Skillfully, we
witness. Having become larger vessels, having developed receptive means through our cultivated attention to beauty, we welcome what we can. We sustain our attention, and translating the
signals through our psyches, transforming them in our bodies, we
shift the degraded form into a chosen gesture. We plant flowers,
make beauty, making beauty in our daily acts. This, I believe, is
Bodhisattva work. This is both receptivity in action and reciprocity. Informed and scented by beauty, duty becomes desire, received as gift, returned as gesture, skillful, generous, and wrapped
evermore in the sensations of beauty running through the body.
By this time, our experience is sensuous, even sexy. We want more.
Iterative. We give more (attention). The beauty (and the beast)
soaks in further. Iterative. We give more (attention and care). Iterative. Now, we're hooked, nourished by the natural magnificence
and mystery of the world. We cannot help but give back. Iterative,
like leaves falling and buds bursting.

Starehe is my favorite Swahili word. It means "to be at ease." It
also translates into *being at home in the world*, a graceful state of body
and mind that arises from reciprocity, from a form of unbroken experience. It begins with listening to whomever the Other may be.
You know, being receptive.

This is my final conclusion, my answer to the conundrum presented by an ever-changing context: To be skillful is to honor the
gift and power of choice. Open or closed, or anything in between,
you choose. For me, the choice is to be either at home in the
world, or not. I choose to be at home in the world. Whenever possible, I choose the unbroken experience. And having tasted it, I
want that seamless and oh-so-sexy feeling so badly, I make it.

Chapter 4

Minding the Relations

Contrast, Qualities, and Patterns

What takes a lifetime to learn is the existence and substance of myriad relationships; it is these relationships, not the things themselves, that ultimately hold the human imagination.

—Barry Lopez

Contrast, Others, and Shifting at the Edge

> Set aside the learned ways of perceiving the world as dead mat-
> ter for your use and see if you can recover again your actual per-
> ception of the world as a community of beings to whom you are
> meaningfully related.
>
> —*Erazim Kohak*

Relationships are the heart of any matter and at the very center of
any human heart's desire. No doubt, it is the heat and hunger of relationships that thrust the human heart into the world, the heart
that makes us look and listen with great desire.

Young Moroccan women traditionally pray to frogs for marriage, as if faithfully throwing their hearts into the world. They
burn incense at the edge of ponds, listening in evening darkness
and doing the "ritual of silent wishes." The only sound they hear is
the croaking of frogs, which they believe are *jinn*, spirits that inhabit the Moroccan landscape. With the intensity of a young
woman's desire for love, and with their senses turned away from
human interaction they are ritually tuned to the spirits.[1]

I am hiding out on the coast of Maine. It is springtime, and the
frogs capture my attention every evening as they begin their fevered
call for mates at dusk. It is the peak of the mating season, and I hear
the frogs begin their reckless chorus as if an alarm is suddenly going
off. I am falling in love with a man who lives across the country, and
for several weeks I also wonder about "the ritual of silent wishes." I
finally give in to my wonder and go to the local frog pond at sunset,
as the chorus begins its nightly and potent serenade.

The pond is hidden between two secret beaches on the end of a wild and windblown finger of land pointing across the Atlantic to Portugal. Wild roses, reeds, grasses, and big patches of poison ivy line the sandy track to the pond. I go there cautiously and quietly, even as a kind of density gathers in my ear. The loud chorus abruptly silences as I near the pond—until I too am still. Then the calling begins again, gathering in loudness and density quickly, as if there is not a moment to spare. A sense of urgency pervades and pulses over the pond. I ask, in silence, why the young Moroccan women pray to frogs for marriage. As I listen, the chorus suddenly shifts into a thousand distinct voices, each voice with its own unique qualities. In response, I think to myself, there are so many voices, so many loves.

I think of learning how to listen to the distinctions, the many different voices now twining through my body. I imagine learning to love the texture of each voice. Then I find myself listening more deeply, believing that there is only one voice that calls for me. I am now listening with my heart, with my heart in my ears. My awareness spills into the early night, now cast out over the pond. My ear adjusts, recalibrating itself to fine differences. I can almost feel the tuning of my ear as I listen to the many voices with a newfound intensity, as if they are calling for my distant love. The twilight becomes even more dense with sound, my heart now calling with the Others, thrown into the wind, the growing darkness, the wild call of *jinn*. I am listening with every pore to a primordial desire, my own desire now reflected in the sound. *I will not forget* the sound. *I will not forget* listening like this, listening with my heart.

But we do forget—the wetlands and frog ponds are filled for parking lots, the voices now gone and forgotten. In our forgetting, the edges of the world are frayed, the seams beginning to tear. Wetlands are filled, forests are flattened, and prairies are plowed for profits. We look the other way, not noticing the loss of lives, the loss of our relations, and just barely noticing the consequent loss of our own experience, of birdsong and the secret places that feed and foster the imagination, the full flowering of our children.

We live in an interdependent world, with every single life in relationship with many other lives, many Others. The field of relations that constitute a human life is endless, and perhaps best described as everything we directly experience. Yet this merely describes our *local* set of relations. And still, this recognition is much beyond our common conception of relationship. We commonly think of relationships in terms of other humans, drastically narrowing the potential field of relations. Consequently, we do not readily see the ways in which the world is woven together, the ways in which one thing influences another, the ways in which we are held within rich webs of more than human relationships. This, then, is why our environmental crisis is essentially a crisis of perception. The fragmented conditions of our world reflect our forgetful blindness, our ears tuned to the human conversation, our eyes tending toward the human creation. But the vaster intelligence of the world is calling on us to hear and truly see the Others. We are being asked to foster a relational way of seeing, to read the signs, to recognize the patterns, and to feel the pulses threading through us. A relational way of seeing the world places us fully within the field of our many relations, sensitive once more to the volume, the width, and depth of being within an animated landscape. From a relational perspective we begin to recognize the many ways in which we are inescapably woven into the world, our actions rippling through webs of relations.

Most of our relationships with the world are either on the edge of consciousness, seen as if from the periphery of our awareness, or entirely invisible. The myriad relationships *around us* are also not always apparent to us. They are not the material things of the world but rather the gestures and processes that exist between the "ten thousand things." They are apparent in the stories that weave the things of the world together, in the way things change, and in the qualities. But many of these unseen relationships are not necessarily hidden. For example, from an ecological science point of view, the presence or absence of indicator species, like frogs, reveals the health or toxicity of the local ecosystem. In this case, the status of the ecosystem is perceived indirectly, through the relationships within the system. For the farmer, cloud patterns seen over several hours, days, and years—from a seasoned perspective—reveal the weather-to-be. The Aral Sea, drained and littered with abandoned fishing boats, reflects eastern Europe's environmental crisis and our collective influence on natural systems. Animal tracks are patterns that tell us who's in the neighborhood and what's going on. The careful observation of the tides teaches us the basics of chaos theory or the way the world is both ordered and unpredictable. Orion, too, is a pattern, a constellation of stars that reminds us of the way of the warrior. Tadpoles, in the act of becoming frogs and watched with the wonder of a child, show us the possibility of transformation, of becoming.

These are all examples of relational ways of seeing, often overlooked and yet relatively visible to us. Whether or not we notice them depends on the focus of our attention, on how deeply we are looking, and on how willing we are to be uncertain, relinquishing the habits of absolutist, dualistic thought. These ways of seeing are associations and patterns representing causal, correlational, symbolic, and metaphoric relationships. They offer perspective, the careful gathering together of numerous experiences into informed ways of seeing, and they reveal the ways in which things and systems influence one another. But my ultimate interest is in perceiving the relationships between our imaginal, intuitive, and sensuous selves and the rest of nature. These relationships with nonhuman nature speak to us most deeply, conversing with our hearts and bodies. In an era hungry for heart and meaning, they remind us of our belonging and purpose within a tremendously vast field of relations, one much larger than the purely human realm. They offer conversation, through the senses, with the powers of the planet—with the forms and forces of nature that are whole and healthy and bigger than ourselves.

A late Arizona afternoon is slipping into evening. The air is silken and golden with light, the sounds soft and seductive. Wanting to savor every bit of this moment between day and night, I hold up my hand at arm's length and count the number of finger widths between the sun and the horizon. Two fingers fit in the space of sky just below the sun, telling me that I have thirty precious minutes to linger next to my favorite high desert creek before the sun sets and I must trundle back to the truck.

I recall a moment when I felt beautiful here (or was it simply *feeling beauty?*). I was lying in the sand, watching young cottonwoods sway in an early spring wind. I could see that they had weathered a nasty storm months before when the creek had swelled over the banks and whole trees had crashed through them. The cottonwoods were littered with broken branches, leaves, and dried mud and were still leaning downstream. After my mother's death I weathered an emotional storm of similar magnitude during the same rough rainy season, and, like myself, I could see that the cottonwoods were survivors. They were also about my age, and I couldn't help but identify with them. But their leafy new beauty was vivid and vibrant, and identification soon slid into a kind of seduction. I was captivated by their bright green grace in the aftermath of such nasty weather. For a moment I was mesmerized, losing myself and gathering up the easy sway of cottonwoods whispering to me in wind. For a moment, I too felt gorgeous and graceful, as if they had revealed beauty secrets to me.

By calling the cottonwoods "like me," by scaling the natural world to that of human concerns, I run the risk of anthropomorphic projection. But in this case, it is just as easy to see myself as "like them," to feel their beauty seeping into me and to imagine my gratitude seeping back through an unseen field of relatedness. This, too, is a relational way of seeing, a reciprocal, unbroken way of feeling, of soaking in and giving back. From the perspective of the sensory body, it is being filled and satiated by light, by sensation, and in return simply giving oneself away.

We court the unbroken experience when we perceive wholeness in the landscape, our view unbroken by fences, roads, or reasoning, or any of the divisive forms that so mark human edifice and artifice. The unbroken experience arises from a stretch of open water, an uncleared forest, an entire day without an agenda to accomplish. A taste of wholeness is the feel of being part of something big going on. It is also the experiential root of integrity, the experience of being rightfully integrated with the Other. It is the seamless experience, the moment of unification while sitting in savanna, the forgetting of one's self. In those moments, the feel of connection and kinship with the Other is salient. We do not question our belonging in the world. Our bodies know that we belong with the Others.

But who is this Other? In a society in which "seamless" is seen as uncomfortably close to codependence, in which we have yet to refine our understanding of the distinction between codependence and interdependence, the Other is often not even a distant relative. By Other—the cat next door, the sister, the lover, the oak grove, the soil and sand, the flower that grows there, the sky—I am referring to that which is in some way different from ourselves and so provides an opportunity for relating. We often miss this, that the Other is an opportunity for relationship—as is apparent, for example, in our conventional forms of power, our collective habits of discrimination. Forms of discrimination (perhaps most often arising from fear) ultimately lack awareness of the qualitative differences, the qualities of the Other. Rather, the Other is often sized up and quantified, objectified, and made distant. We thus do not see the Other clearly, missing the distinctions and differences that show us what is possible for a life. Most fundamentally, difference is a manifestation of potential, the many spices of life displayed for us, reminding us of our own potential. This is a fundamental recognition of relatedness. But without a more relational perspective, we rarely recognize such patterns and their lessons.

Given the prevalence of dualism—exemplified by reductionism as the scientific or culturally sanctified method for knowing truth, by our collective focus on objects, and by our self-obsession—differences are readily amplified in the modern mind. John Dewey and Gregory Bateson, both very concerned with the effects of dualism on modern thought, agreed that "one dualism leads to another."[2] In the calculus of the modern psyche, differences easily become confused with divisiveness, revealing a sensibility infused with a kind of insistent and unconscious absolutism. We all too easily avoid the gray area, the inherently shifting qualities within the realm of relating. Rather, we are supposed to know, to have answers, clipped, concise, and right. We may be right, but something's wrong. Without an eye toward the Others, we find ourselves alienated, somehow fragmented. It is as if we have missed the true signs, the look and feel of relatedness of greater wholeness.

The experience born of giant parking lots and days divided into fifty-minute hours are, most fundamentally, "broken experiences." By "broken experience" I am referring to the range of experiences that fragment, that shatter wholeness and connectivity, that disrupt the integrity of living systems. I am referring to, for example, the bells that mark the end of math class, the twenty-minute lunch, the fifteen-minute recess. Even as young children we are conditioned to the broken experience. We learn the transience that alienates us from place and community, just as we learn the disrespect and abuse that demand a chronic vigilance and defense against the unknown Other. We learn to stay within fences and to build material and metaphoric fences that divide and ultimately sabotage seamless experience. The broken experiences are unnatural and fundamentally discontinuous. They are nonparticipatory and nonreciprocal in relation to everything on the planet except modern human creation and human material need. We are left with a hunger for wholeness, and yet unconsciously and habitually, we continue to see distinctions as absolute and often extreme differences.

Within the traditional Dineh (Navajo) worldview, distinctions are bridges, ways to identify connections rather than amplify differences. Distinctions are viewed as pairs: Night is paired with day, man with woman, east with west. East signifies dawn and new beginnings. West is the place of sunset, the shift into darkness and inward reflection. Conceived as paired powers, east-west signifies the union of initiatory, activating energy with that of a reflective and inward-looking ability.[3] In observing the daily movement of sun across sky, this union is recalled and amplified by Talking God in the East and Calling God in the West. Together, in myth and ritual, Talking God and Calling God initiate and support human intention and activity. This is a relational view of reality. It places opposites on the same spectrum. Conceptualizing paired yet polar energies cultivates a sensibility of balance and unity and influences Dineh reality. The world is seen in unified terms—body with mind, earth with sky, matter with spirit, the immanent with the transcendent.[4]

Relational ways of seeing must begin with an appreciation of the Others. We must look outward, beyond ourselves, offering them our attention. But again, who are these Others? They are both signifiers of difference and the source of relatedness. From a profoundly relational perspective, they are woven into who we are. In *The Others: How Animals Made Us Human*, Paul Shepard, an environmental philosopher, asks how the Others have made us uniquely human. His answer draws upon a profound recognition of our history, of what he calls our "ecological heritage."

We have coevolved with natural environments, the places that constituted our homes, for some 2 million years. We are old; our brains for 90-some percent of human history, up until the last ten thousand years or so, have been most engaged in hunting and gathering. In the act of hunting, the human mind evolved the capacity to read and memorize tracks, pulling patterns from the background. "The human mind came into existence tracking," says Shepard. "Mind would be the child of the hunt." As patterned objects, tracks were spatially encoded, mapped both in the outer landscape and in the mind, thus providing us with "all kinds of data." We learned the nuanced signs of different animals and their behaviors; we learned the pairing of particular animals with particular places; and we learned to orient ourselves in space. And because tracks distanced predator from prey, they were a visible source for understanding the past and the future. They were good grist for thought.

Consciousness thus evolved within the dynamic dance between predator and prey: Our minds were led into pattern recognition, and through seeing a footprint in sand and soil, we developed the capacity to both remember the past and imagine the future. In the human analog of deep time, the relationships between ourselves and the animals that fed us have crafted a unique human consciousness. The animals thus "made us who we are." The Other is woven into the form and function of our brains and bodies, our whole way of being in the world. Tribal rituals, now rapidly fading from view, honor this relationship with the Other, offering respect and the kind of attentiveness inherent, for example, in totemic belief systems. The animals are recognized as guides, deeply informing the human mind with observations and stories that outline the cunning quality of coyote, the grace of gazelle. Beyond gaining the ability to see, remember, and imagine, we learned to be tricksters—playing jokes like coyote—or to be graceful and swift.

The Others are our contrasting kin, both different from us and intimately woven into who we are. This is perhaps a radical perspective for the modern Western mind. Our identities do not usually include a sense of such woven-in intimacy with the Other. If we entertain this perspective, however, it has the potential to nudge us out of our divisive tendencies. With such a perspective we begin to realize that the things we see—the sunrise and sunset, the springtime bloom, the cycles and patterns that guide our survival and satiation—have crafted and shaped the very eyes with which we perceive the world. And with the thought of contrasting kin, we more easily perceive the fundamental pulse of the planet, the ultimate nature of both polarity and synthesis, the "moving signature of the universe,"[5] the continuous ebbing and flowing inherent in any relationship. We recognize that there is no one way, no absolute right or wrong, no stasis, and no final word. Our existence is never a done deal. As such, the Others lead us, as if still guiding us, into a non-dual recognition of the world.

It is tempting to make dualism out to be the true culprit of the modern mess. It is obvious that divisiveness and separation as a habit of mind have been recapitulated throughout our culture to such an extent that we mistake stasis for completion and completion for good and right. We mistake consumerism, the endless accumulation of things, for the saving of our lost souls; we bury our lost sensuality in the thin guise of rationalism ad nauseam; and we limit truth to right or wrong. But to say that non-dualism is the only way to redemption is dualistic in itself. The postmodern mind ought to know better.

A relational perspective encompasses both difference and unity as fundamental realities. The human visual system models this at the very instant that it absorbs light. To encode a signal, the first neuron in the system requires the signals from two neighboring receptors, for it is the *ratio* of their signals that is needed to further encode the input. This relationship—this ratio—is the essential code that allows us to see in the first place. It is both a difference signal (coupled with different intensities, or contrast, in the visual field) and a unified relationship. Without the visible contrast, without the encoded ratio made possible by difference in the visual field, the perceived world quickly fades to a dull gray.[6]

Contrast is the most fundamental of the visible relations. It is the signal that one thing is in relation to the next. It appears as the sharp, shiny edge of black rock against a bright blue sky. It is the edge between dark and light, between round red berries and brilliant green leaves. It is the horizon, the edge of night, the edge of shadow cast by dunes and ripples in sand, the line that distinguishes this from that. It is a thin line of branch across a full moon, a star against the night sky. It is the edge of water on shore, lines of volcanic infusion through an expanse of rock, and the difference between faith and doubt. In all instances, contrast marks an edge.

Seen by the human eye an edge is not perceived as simply an edge. It is a zone in which the signal is amplified. In literal terms, this amplification is a perceptual phenomenon that occurs when

visual fields of different lightness or color are adjacent, or in relation to one another. The change at the edge, the shift from one
thing to the next, is enhanced or amplified as the signals pass
through the eye. In visual science, the amplification of the signal
is known as the contrast effect. A common example is the bright
band of light—the Mach band—that runs along the edge of sky
that touches a ridgeline, just after sunset, and the corresponding
dark band appearing on the mountain side of the horizon, just below the ridgeline. But contrast effects are also apparent in the intensity of a cat call I hear, shattering an otherwise silent night, and
in the feeling of stepping free of gravity as I take off a heavy pack.
Contrast reveals my own cultural conditioning as I immerse myself
in another cultural context and intensifies the pleasure I feel after
ceasing the proverbial "banging of one's head against the wall." In
all of these cases, we feel the difference with an added dose of sensation—the amplified effects of contrasting signals. It is as if we
experience the difference with a little jolt at the edge.

This amplification is a perceptual effect, occurring within the
sensory systems as opposed to "out there," in the physical world. A
physical measurement of the light intensity does not reveal an increase in the edge of sky or a decrease on the dark mountain side
of the edge. The explanation for the perceived amplification is lateral inhibition, or the way in which the receptors and neurons are
wired within the retina. This translation of the visual signal is a

hard-wired effect and is, quite notably, one way in which human perception alters objective reality, the kind measured by physicists and physical instruments. In this case, the translation illustrates the fact that even two-dimensional visual fields, such as black-and-white stripes, change in appearance by virtue of being in relationship. But as John Berger, author of *Ways of Seeing*, points out, "Visually, *everything* is interdependent."

Extrapolating this principle to include the things of the world, our perception presumably differs when we see objects in relationship, as opposed to within a kind of perceptual isolation. This suggests that an inclusive, contextual view of the world differs in appearance from one consisting of isolated, independent objects. With a little reflection, we know that the context of an object reveals its significance, what it signifies—its true sign. In other words, the true meaning of a thing is seen only with reference to its relationships—and, the relationships are most readily apparent by attending to the edges between things, where the differences are amplified.

The edge is where one thing becomes the next. The edge is also shared, although it seems to belong to one thing more than the other, to the mountain more than the sky, the self more than the Other. We tend to see the edge as the property of the figure we are attending to. But any edge also delineates the ground—the background, the negative space, or the unattended form. It marks what is on the *other side* of our attention, of our current reality, signifying that the world is more. At the edge of our consciousness, for example, the mind teases us, suggesting that the other side is endless, fathomless. The edge between land and sea, like other

ecological edge effects, is teeming with life, with abundance, as
species stretch between ecological zones, as if the world is more,
always more!

How we perceive either side of an edge shifts our perception of
reality. This is illustrated by the Rubin Vase, a classic illustration
used in perceptual psychology textbooks to demonstrate figure-
ground reversals. But we might also consider it a demonstration of
perceptual flexibility, of our ability to see the world anew, with a
new perspective. When the edges denote a center white vase, the
black profiles become the background. The figure, or vase, is then
reversed—or replaced by the ground—when the edges are seen as
delineating two identical profiles. In simplified terms, this is a form
of shape-shifting accomplished by virtue of our attentional focus.
It requires the ability to "give substance" to either side of an edge.
In some small way shifting our attention like this reverses the
world.

Legitimizing and practicing a relational way of seeing expose
the possibilities and shiftiness of the world. Nothing stays the
same. The world becomes dynamic—the edges amplified, calling
for our attention and signifying a change. We more readily see the
ways in which things influence each other, the systemic perspectives now shifting into the foreground like figure-ground reversals.

The world becomes alive, animated, and our sensory selves feel the tug, the pull to engage, to step into the picture as participants. We find ourselves responding, feeling and being "in relationship with" the things around us. We thus participate with more than the merely human world. In relationship, at the permeable edge between ourselves and the Other, we become more than our independent selves. We are now influenced by our experience, our participation with the sun and wind and tide. We are influenced by a change of season seeping into our bones, and into our souls. We are reflective in the dark of winter, flirtatious in springtime light. As women, we are influenced by the tug of moon moving through our bodies and bringing blood to the surface. To perceive this, our minds and bodies must open, must be desirous enough to relinquish the habitual and protective layers, to loosen the edges. We must surrender ourselves—or, more precisely, our *ideas* about ourselves.

Buddha's central insight and teaching was the recognition of profound interdependence. In Buddhist terms the fact of the relational world is referred to as dependent co-arising, in which the world is seen to arise continuously and in the context of relationship. Buddha's teachings include a phrase that refers to a way of being that is necessary for perceiving the nearly invisible and yet ubiquitous fact of relationship. The phrase is *yoniso manasikara.* Manasikara comes from a verb meaning both "to ponder" and "to take heart." Most essentially, it refers to a form of attentive pondering, a specific, heartfelt form of attention. The pondering is qualified by yoniso, related to the word *yoni*, the Sanskrit word for womb. By extension, yoni also means "origin," "way of being born," or "matrix."[7] So yoniso manasikara refers to a form of attention in which we give ourselves over to the Other, surrendering ourselves to the matrix, being reborn into a newly appreciated and interdependent sense of self. By surrendering our independent, egoic sense of self, the relational world is naturally revealed.

Buddha's teaching regarding the perception of interdependence was that we offer ourselves to that which we see. The offering took the form of "pondering with heart," or looking long enough to truly see and value the Other, perhaps looking with "love eyes." In the process we surrender the ego, perceiving ourselves in relationship, now woven into the matrix. Such a perceptual shift is a reversal of the world, the ground of our being—the relational field within which we are embedded—shifting into primary awareness. This is at least part of what our ecological crisis is asking of us.

The full-moon tide is slipping into the estuary. It is fingering its way through spiky young marsh grass, over saturated banks, and across stretches of marshland. The long blond strands of last year's grass float on a silken surface, swaying in, then out, then in again with every subtle surge of tide. Wet, yielding, and thriving, the grasses grow thick and luscious green with the flood of salt water. I too yield, the tide pulsing in me over many days and weeks, my blood flowing with the big moon tide.

I watch the sea become land, become sea. Where is the edge? Where does one end and the other begin? Grass, mud, and tide ebb, flow, and pulse together, the sensations now seamless and twined together with the rhythm of my heart. With the subtle surge of tide, I too grow thick and luscious.

Patterns of Relationships

> A good pattern is but an enhancement of what is true. Pattern is
> the transmitter of beauty.
>
> —*Soetsu Yanagi, philosopher and art historian*

My eye follows an amber line down the edge of a densely patterned Persian rug, now turning a corner and becoming burgundy.
It shifts into a woven form, then twists itself into something like a
star. Then starbursts carry my eye into a new pattern I cannot
name. I am dazzled by color and form leading me into a way of
seeing that by its very nature is emergent, always hinting at something to come. As my eye continues to wander across shifting
patterns, the puzzle pieces rearranging themselves into new configurations, I find myself wondering, Who wove this carpet? What
was the quality and complexity of her mind? *How in the world* did
she *see* the world? The edge of my mind shimmers with all sorts of
possibilities.

Patterns are ordered and emergent relationships. Visible patterns
are rhythms in space. Diane Ackerman claims that patterns can be
defined as an instance times at least three. "Once is an instance," says
Ackerman. "Twice may be an accident. But three or more times
makes a pattern." She adds, "Few things are as beautiful to look at as
a ripple, a spiral, or a rosette. They are visually succulent. The mind
savors them. . . . We crave them."[8] According to Gregory Bateson,
patterns are "the bones of the universe." They hold together and order the world, giving rise to wholeness and identity.

As the bones of the universe, patterns reflect that which is known, stable, and orderly. Spirals, starbursts, meanders, and ripples appear repeatedly in flowers, in streambeds, and in our minds. We see these repetitions of form across many dimensions of reality, as if the patterns of the world were heading off into every domain, radiating from our point of view. Within a maple leaf we see a repetition of lines forming an adaptive geometry of veins and revealing a pattern that is shared among leaves. Simultaneously we see the unique lobed pattern of maple leaf. We recognize the five-

lobed pattern of Bigleaf maple leaves as we walk along a wet forest edge near water showing us that the world is ordered and held together. Constant, repeating elements remind us that we live in a reliable universe, as if the world were offering us the reassurance of order in what would be an otherwise unknown and chaotic universe. We see, quite literally, that much of the world is something we can count on.

Patterns order the world and yet simultaneously call us beyond ourselves, showing us yet another form, another quality, another way to shape a life. Like the patterns arising in a complex woven rug, like the endless permutations of snowflakes, and like the individual variations found within a single species, patterns shift and change, continuously revealing both the connectivity of the world and the previously unseen. The confluence of the repetitive, constant quality of a pattern and its emergent and transformative qualities reveals the nature of adaptation, showing us how life shapes itself to, or with, its environment.

In the natural world, repeating patterns shift to reflect the specifics of time and place, the context, the set of relationships within which it is embedded. For example, leaves growing in shade and relatively moist air tend to be more round and less differentiated than leaves growing in relatively dry microclimates with much light exposure. Even within a single plant, the leaf pattern found close to the moist ground and shaded by the rest of the plant is round relative to the pattern of leaves found near the top of the plant, where there is less moisture and more light. This shifting leaf pattern signifies a relationship between the characteristics of the landscape and the plant. Mark Riegner, an ecologist by training and a phenomenologist at heart, calls this "form language." "Organism and landscape form an inseparable unity,"[9] he says. Seeing the shifting pattern reveals not only the context—the plant's environment—but also "the pattern that connects" the relationship. Such shifting patterns are doorways to more inclusive patterns, to seeing more holistically.

The spired shape of fir and cedar and pine is a "pattern that connects."[10] The spire connects Douglas fir with Grand fir and with Redwood. If we stand in a moment of contemplation, with even a hint of reverence, the spire also connects to our own patterned notions of heaven, of invisible powers, and perhaps even to our feelings of aspiration. The pattern of our thoughts and aspirations is embodied and made visible by the spire of a redwood. In contemplating "redwood," we *feel* the vision, the image in our bodies, as a rise, as an uplift, and as reverence.

Patterns that connect reach into the unseen space between the things of the world, revealing constellations like "the four-leggeds" or "the winged ones." We perceive that things belong together. I see myself *belonging* with the spire, the reaching and rootedness of Douglas fir, Grand fir, Cedar, Ponderosa pine, and Redwood, the great evergreens of this land. Like them, I too stand on the surface of the earth, mediating between the material world and the realm of the sky gods. I recall (I invoke) the Tree of Life, its uppermost branches reaching into the home of the gods, the roots spreading laterally, just below the surface of the known world. I remember (I imagine) the Omphalos, the cosmic axis or *axis mundi*. The Omphalos is the mythic center point of a sacred circle within which the shaman traditionally moved between the underground world of soul and the powerful sky realm—through the world of earthly concerns. Across many cultures, marking the Omphalos is the first act of geomancy, or sacred geography. This act declares that the sacred center of the universe is *here*, and that the cardinal directions extend from the "point of cognition," the place where one stands. The Tree of Life, a tent pole, sacred pits, and ladders—all symbols of the axis mundi—place the mind fully at the center and fully in the landscape. I remember (as if feeling such memory in my body) that the Omphalos is a sacred center where the potential of mind, the possibility of transcendence along a vertical axis, intersects the horizontal axis, that of the ground, the earthly realm of soul.[11]

A large sycamore stands before me. I see roots fingering their way through river rock and digging and spreading into the edge of a creek. I see branches that carry my wandering eyes outward and upward into the sky, my heart carried with them. The pattern of thought and reverence that connects me to the tree is just one step beyond its visible branching pattern. And as I let my eyes leave a branch and dive into the sky, I am pulled, even tugged, more deeply into the universe. It is as if my mind has suddenly grown larger.

Patterns arise naturally when our eye follows an edge. Our eye wanders up a trunk and alights on a branch, the edge now leading the eye on to the next branch, and the next, then to the pattern of branching. Our vision thus transcends what we initially see. We see the patterning, recognizing the emergent form, the larger understanding. Our eyes are led from one branch to the next, to a metaphoric recognition of the path taken or not taken. We remember (as if the tree holds the memory for us) the metaphoric "Garden of Forking Paths," and the branches become a visible reference to the power of choice, the path chosen or not chosen.

Patterns are poetically dense. The spiral, for example, is the gesture made by sunflower and pine cone, chambered nautilus and rose. It is the pattern of the hero's journey, the celebrated return, and the pattern of my growth, ever larger and more inclusive. Pat-

terns lead the eye through the rational mind and into an imaginative mode, into mythic and mytho-poetic ways of seeing.

Patterns can be found both in the sensible world—in the form
of a maple leaf—and just below the surface of visibility, like a
shimmery intuition coalescing into a form of recognition. In the
invisible realm, patterns of behavior, for example, rise into consciousness with gathering memories, gestures, and signs, offering
form to what has been unconscious, unknown. The initial recognition of such patterns stretches and expands our consciousness,
shifting our view of the world, adding to our repertoire of recognition, and teasing us into a deeper level of knowing both ourselves and the world.

Metaphor is a particular form of patterning. When we perceive
metaphorically, we note a likeness, a similarity in form or organization, a kind of symmetry between patterns. My friend Robert

Greenway—an early ecopsychologist and original wilderness guide—describes metaphor as "when the symbol takes a mate." He adds that its mystery is found in the way that it "reaches out" for the Other, as if a metaphoric likeness between things inspires magnetism. For Gregory Bateson, metaphoric refers to the language of nature, the language of relationships, and of story. Linda Olds, author of *Metaphors of Interrelatedness*, writes, "Metaphor directs attention to similarity in structure or organization between realms or events."[12]

From the perspective of psyche, metaphor is the root of relatedness, the linking of things, events, and constellations of experience. When our senses are well tuned to affinities, metaphoric ways of seeing become deepened into ways of knowing. Metaphor is a way of being informed by the Other, recognizing and acknowledging the likeness with language. We say, "a bed of moss," a "stream of consciousness," and a "flood of insight." We say, "My mind meanders like a river," or "My heart pulses like tide." We say that the sparkle in her eye is like light on water, and her voice is liquid. Metaphor is a very short story that stitches the world together, weaving similarities with recognition and language. For Bateson, metaphor is the very ground of kinship.

Metaphoric perception requires us to learn a language, the language of nature, of all the relations, of relatedness. It is a language cast in images. "The language of the rock is its ability to show the seeker images," says Leslie Gray, a shamanic counselor.[13] It is a language arising from patterns that connect, a language in which Mountain *displays* a form of solidity and speaks of—by showing us—massive presence and the possibility of perfectly still meditation.

Patterns and metaphors are all over the place. Patterns are the very essence of life, says Fritjof Capra,[14] and metaphors are ordinary.[15] Seeing them "all over the place," however, is a matter of shifting our attention. When we redirect our attention from the analytic mind to an intuitive form of sensing, we "see into," as Goethe said. We develop "organs of perception" by recognizing the mountain for what it is, by letting it reveal itself, its essence with respect to our intuitive insight. We cultivate a receptive mode of consciousness that allows us to read the language of the phenomenon. The language is nonverbal, spacious, and born of how things really are. Allowing phenomena to speak for themselves, turning "toward the world as it is experienced in its felt immediacy,"[16] is the essence of phenomenology: According to Henri Bortoft, "The phenomenon becomes its own explanation. It discloses itself in terms of itself and thereby becomes self-explanatory."[17] The analytic mind can take a walk.

To see in this way, perceiving patterns and metaphors "all over the place," is presumably the gift in Goethe's method for developing organs of perception. His method included perceiving the "intensive depth of the phenomenon," which was gained by gathering numerous views of the same thing. "Gaining perspective" arises from multiple perceptions and angles, from the coalescing of experience. Observation from a variety of literal and figurative angles, or with many return visits, deepens our appreciation for what is seen. Perspective adds depth or dimensionality to any perceptual moment.

Goethe's way of science, of revealing truth, was to develop a perspective that reveals the pattern that connects, that sees "unity in multiplicity." The process gathers together the qualities that are shared between phenomena and yet preserves their distinctions.

Attention to the patterning between things clarifies the shared form or organization and, contrary to abstraction or generalization, does not minimize the distinctions. Generalization is the melting pot approach, our first social science lesson. It was thought to be a great lesson by the educators of my childhood, but an entire generation seems to have lost a perspective born of diversity, a recognition that differences make the relational field apparent and render depth. Without a ready appreciation of differences, we see little of the Other—we do not perceive the Other in depth. As an alternative to generalization, Goethe's practiced view was both differentiated and holistic. According to Bortoft, seeing in this way is a "higher cognitive function" than generalizing.[18] Bortoft further claims that a particular quality of attention is needed in order to see in this way: "What is really needed here is the cultivation of a new habit, a different quality of attention, which sees things comprehensively, not selectively." This idea is particularly interesting in the context of the conventional psychological definition of attention: "the enhancement of *selected* sensory information." The conventional emphasis is on *selecting* a portion of the visible field.

For Bortoft, and Goethe before him, the shift in attention necessary to see the pattern that connects, and to see the intensive depth of a thing, was toward the intuitive mind. This is not about either zeroing in on a selected portion of the visual field or spreading one's attention across the field. It could be either or both. It is shifting one's attention to another perceptual mode, as if seeing from within the framework of a new paradigm. From this perspective, the message is the medium. The message in the mountain is to be found in the direct experience of Mountain—not in the analysis, but in the phenomenological, hermeneutical, and intuitive apprehension of it. The hermeneutics of perception is a cyclical, reciprocal process of seeing *into* the phenomenon and *into* one's practiced imagination. It is allowing the phenomenon to show itself in a moment of "pondering with heart."

A heightened perception of patterning reanimates the world with meaning, with signs and forms that coalesce and belong to one another, the world now woven into greater wholeness. Soetsu Yanagi, a Japanese philosopher and art historian, claims that the value of patterns lies in their "vitality, in their transformative powers and their ability to metamorphose, symbolically, wisdom into its highest order." He adds that patterns "provide unlimited scope for the imagination."[19]

For traditional hunting peoples, the animals represent complex patterns of behavior, displaying qualities and possibilities for a life. According to Barry Lopez, hunting people hold animals in "high totemic regard" not merely because they are good to eat but because they are "good to think." The animals are "good to imagine." Within earth-based cultures, the animals are said to have certain qualities, certain powers. These qualities are directly observed, the animals seen and contemplated again and again. For the Kalahari Bushmen, the qualities are coalesced into stories and messages. There are teaching stories about every animal—the hippopotamus, the hare, the porcupine, the praying mantis. The praying mantis, for example, is the Bushmen's image of God. The mantis is observed as a small egg transforming itself into larvae and worm and then further metamorphosing into a winged insect that is able to hunt and devour other insects. It sits in a position of great reverence, as if praying. It has a face that is, in the words of Lawrence

Van Der Post, "curiously human, almost like that of the Bushman himself." And in the act of mating, the male is devoured by the female. In this way, the praying mantis is a "comprehensive symbol" of birth and rebirth, portraying the Bushman's own spiritual development.[20]

Totemism is a dynamic understanding of the landscape. It is based on careful and repeated observation of the totem—Raven, Salmon, or Seal—infused with memory and the stories shared by a culture. Raven becomes a constellation of abilities and gifts, a form of grace and unmistakable presence—gliding, diving, and swooping; conspicuous, black, abrupt, and loud in an open sky. For many of us "two-leggeds," Raven is a messenger. Who could not take notice of him? He often interrupts us, punctuating our human conversations with his raucous cries as if he has something essential to say. He mutters, chatters, chortles, gobbles, croaks, squawks, screeches, and banters. He clowns around, flying upside down or pacing back and forth on a slick rock ledge, as if laughing at the world, his hands tucked behind his back. Don Juan, Carlos Castenada's mentor, noticed that Raven never competes, apparently more interested in simple pleasure than in success or victory. The way of Raven, we might imagine, is both free and outspoken, unattached yet with plenty to say. The apprentice to Raven learns this, the patterns of behavior that are Raven-like, the way of Raven. As an apprentice to this totem, she, too, is a messenger and is expected to have something to say, to offer to her community. And presumably, the community holds her to it. Pattern perception, as a learned skill inspired by the natural world, thus becomes both a form of education and the foundation for a sense of belonging, of having something to offer one's community.

When we moderns attend to the patterned world, we ready our neural circuits for making associations that extend much beyond linear causality. Beyond our learned presumption that *this causes that*—and well beyond the dualistic perception of *this or that*—we become perceptually prepared to see webs and spirals, to see the things of the world gather together and emerge into new constellations. As we become more accustomed to seeing patterns emerge before our eyes, we more readily perceive the less visible patterning in our own lives. We might call this a "transfer of learning," or developing a good intuition. Intuition, we now realize, is the edge between unconscious and conscious pattern recognition. It is both the ability to perceive subthreshold patterns *and* the emergence of a pattern into consciousness. Aha!, we say, suddenly grasping what has been just below the surface of our recognition all along. And so with true reciprocity—our attention given over to the patterning to be found in the relationships between the things of the world—our perception, more seamless now, more ready to perceive the relationships, reflects itself back into our eyes, penetrating us. *We see ourselves* mirrored and potentiated in the myriad patterns of leaf and limb, of animal totems, of spirals, the ripples and the meanders that carry us downstream. Our own true, natural, organic selves are revealed. We see our own patterns rise to the surface as we find ourselves meandering among the relations, as we begin to *get it*; to see and understand that we too are natural, that we too are truly of the earth.

Pattern perception is an awakened sense of both emergence—the gathering together of form—and stability, the repetition of form among the ten thousand things. If Paul Shepard's theory on the evolution of mind is correct, this is just what our attention is made for—both detecting the novel pattern and discriminating the subtle shift, the nuance in what is deeply familiar to us. And so the world, with the force of coevolution well behind it, calls us out, tugs at our attention with the shifting forms of patterns. Our sensory selves, the long-evolved perceptual self, responds. This is the self arisen from some 2 million years of marriage with the cycles of seasons, of hunting, feasting and celebrating, of watching Moon, feeling tides and the rise of River. Our senses have evolved in constant relationship with the Other. With such a deep history, with our senses shaped by this very world, we carry within us a particular sensory profile, one that easily follows the contours of the land, that is born to read the signs, that looks for weather patterns and for the signs that point to a deeper existence. We look for the view that satisfies the senses, that is sensuous and rich with metaphor. We hunger from the core of our sensory bodies, from our sensuality, for such a tug, such a wealth of experience.

The world is made richer and denser by our patterned visions, by seeing the interpenetration of the relations, touching and coalescing, and by ourselves, the viewers, being pulled into the ordering of form and color. We are aroused by the tease of almost knowing, of almost being able to say what it is that we see and recognize. We crave patterns. And although I often cannot fully name or describe the gathering of likeness, the repetition of form, the undulations in water, the spiraling of eddies, the meanders of my mind, I am dazzled.

Forms and Forces: Natural Archetypes

> What's essential is a lake in the mind. . . .
> —*Deena Metzger*

An archetype is a constellation of qualities, a pattern of patterns. It is a grand metaphor, a blueprint that codifies universal human qualities, that speaks a pattern language. James Hillman describes archetypes as "the deepest patterns of psychic functioning, the roots of the soul governing the perspectives we have of ourselves and the world." He notes, too, that archetypes "tend to be metaphors rather than things" and that they are represented by images. In essence, archetypes are what the imagination shows us about "the basic nature and structure of the soul."[21]

Archetypal images are held in both the mind and the sensible world. As constellations of patterns, they gather into forms of revelation. Like less complex patterns, archetypes help us to reveal what rests below the surface of our awareness, the yet-unseen co-alescing into visibility like rainbows, the image becoming more saturated, and more clarified. We begin to recognize the patterns that make up the way of the Warrior, or the archetypal Mother, or the Heroine's journey.

Despite varying degrees of visibility, all archetypal images carry a kind of psychological weight or significance. Robert Johnson, a Jungian psychologist, suggests that the potency of an archetype can be found in its emergent quality: "When we encounter the images of the archetypes we always feel *the power that has been shaped by the image. . . .* We can feel the archetype as a charge of energy."[22] Regarding the power of archetypes, Hillman says, "One thing is absolutely essential to the notion of archetypes: their emotional possessive effect, their bedazzlement of consciousness."[23]

According to Carl Jung, the archetypes appear symbolically in dreams, mythic stories, religious icons and rites, cultural patterns, and "all products of the human imagination, such as literature and art."[24] Jung was the first to emphasize the purely psychological archetype, the patterning that preexists in the collective psyche of all humans. But the idea of archetypes is an ancient one. Plato's "ideal forms," the patterns existing in the pure mind, are in some respects identical to Jung's view of the archetypes. Plato's "Ideals," however, were distinctly inspired by images within the world. For example, horse inspired "horsiness," the Ideal of horse. Over the course of history, we seem to have forgotten the role of the visible, natural world in inspiring archetypal forms of understanding. It is this forgotten detail that has turned our gaze from the power and potency of the land, and thus from the soul of the world.

The Dineh (Navajo) mind is inextricably tied to the land. In describing the relocation of thousands of Dineh following a government order in 1974, Edward Casey, in *Getting Back into Place*, also portrays the heart-wrenching implications of relocation. He calls the experience of displacement from one's ancestral lands, "place pathology," and in the case of the Dineh, likens it to a loss of self.[25] A recent *Christian Science Monitor* article described the current pressure to relocate more Dineh families—125 or so—so that the Peabody Coal Company may continue mining with "shovels as

big as buses." The article ends with the words of Roberta Black-
goat, founder of the Sovereign Dineh Nation. She says, "If we lose
this land, we lose our souls, our reason for being."[26]

The Dineh sense of self is archetypal in dimension. For them,
the self is traditionally understood and made visible in relation to
the land, and specifically in relation to the four cardinal directions.
With every 90-degree turn, with every new direction, the world
appears wholly new, and a different story emerges from the landscape. For the Dineh, the cardinal directions are reflected in Mount
Blanca to the East, Mount Taylor to the South, the San Francisco
Peaks in the West, and Hesperus Peak to the North. They appear
as significant reflections for one's self, seen with each upward
glance cast into the distance, with an eye toward the horizon.

The encircling mountains are archetypes, each constellating
the power of one of the cardinal directions, of Mountain, and of
the spirits that dwell within them. The cardinal directions carry
particular phenomenological powers—the power to begin again,
for example, as seen with each new sunrise in the East—and mark
life processes, the patterns of maturation. The South reminds us of
the child, fresh, playful, and concerned with material things. The
West marks adolescence, the time to question one's self, to become introspective, to go into the darkness like the setting sun. In
the West, one learns the way of the shadow, and the way of soul.
The force of the West surfaces in adolescence, the desire for deepened experience, for a deepened awareness, a transformation of
one's self. One drops below the horizon, learning to see in the dark,
and to persevere. White Corn Boy and Yellow Corn Girl are ado-
lescents dwelling in the West, within the San Francisco Peaks.

The Dineh understanding of self—mapped onto the cardinal

directions, onto the circle of one's place and placement—is rich with archetypal dimensionality and force. One's self is aligned with larger forms and forces and is made visible through the phenomena, through the patterning held by the land. The patterns are remembered by the land, held in the vision of Mountain, a massive white peak arising from a golden plain, arising like a mystery and glowing with liquid light spilling into every direction. Where did such a mountain come from? It must be the very edge of the world. There must be power there.

There must be power there, the power of perspective, the multiple views of the indwelling spirits of Mountain, the ones with the widest view, the land stretched out below them. They are the gods of an ecological perspective. They watch over the watershed, the source of all life below. Their power is in the numinous and sudden rise of earth, the uplift and the phenomenological desire to go there, to get closer to Mountain. In the traditional Dineh view, the four sacred mountains define the horizon, encompassing one's world, encircling the self. Together these mountains form a geomantic map of the world, in which the spirits, the world soul, and the self exist together, all with reference to place, one's homeland.[27] In a geomantic universe, the land is crowded with archetypes and with the wonder and power of the human imagination.

What I call "landscape archetypes" carry a different order of psychological potency or depth than those residing solely as constellations of patterns within the human mind. The difference can be found in the number and complexity of the patterns that comprise or constellate landscape archetypes. The landscape holds a large number of fundamental patterns for us. It "has memory," as the earth-based peoples would say. The landscape has both memory and ancestral wisdom and holds these for us in the image, in the visible patterns. Our minds carry the other half of the archetypal equation, our memories holding perspective and history, the stories of our tradition. Because we do not have to carry the entire archetype in our minds, there is more that can be held, more potential depth in the constellation.

Landscape archetypes are forms and forces of the natural world that simultaneously reveal to us more of the world and more of ourselves. Like the purely human-created archetypes that are familiar to the modern mind, they are patterns that "bedazzle consciousness" and stretch our imaginations. Natural archetypes are highly visible forms that resonate with our deeper intuitions and understanding, pulling the unseen within us into greater consciousness, reminding us of the larger, fundamental patterns that govern our lives.

Rainbow is such an archetypal form. Across all times and cultures, rainbows have certainly "bedazzled consciousness," pulling our imagination right out of our heads in wonder. A rainbow shimmers its way into existence, a mythic vision coalescing before our eyes. We imagine a pot of gold at the end of it and wonder just where this ecstatic arc of color might touch the earth. It must be magical there, where the ground we walk on is suddenly saturated by translucent color, suddenly touched by the gods. We long for the treasure, the pot of gold, the enchantment of the world. Of course, we never get to the end of the rainbow, and that too is part of the message. We are simply beckoned, called out of our mundane selves—if only for a moment—by a great mysterious band of the most magnificent colors arcing across the sky, a rainbow bridge to the gods.

Perceiving the archetypes is a process of seeing into. It is contemplative, and an offering of one's attention. It is, once again, re-specting, looking *again* and *again*. We see into the fact of something—a mountain, a rainbow, a cave—to its inherent qualities. In Water we see the nature of flow, the power of erosion, a patient form of shape-shifting. We see that we may never step into the same river twice. In Water, we see the capacity to take any shape, to be utterly still, like a vast, clear mirror, or to be utterly wild, windblown, and fierce. We watch Water be pulled and tugged by Moon, responsive to the invisible forces between bodies. In both the clumsy walk of a swan and the grace of Swan in Water, we see what it means to be in and out of one's element. We understand utter subjectivity when we see Fish in Water, their only element, their single perspective of the world. Water, we see, speaks to us universally. It is a shrine, a calling, a catalyst for understanding the Tao. "Learn to be water, let the moon strum your belly, the planets beckon and tug," says the poet Morton Marcus.

By fully seeing *where we are*, archetypal forms in the land show us *who we are*, in the most essential, most human of ways. They help us to recover our natural selves, rediscovering and disclosing patterns long buried in the psyche. As if we need hints to recall our own deeper understanding, archetypal forms remind us of our human capacity, of an expanded and exalted understanding of ourselves, of who we can be. By touching and tasting, sniffing and looking again and again, we become tied to the earth, enlarged and wedded by the gift of having a body, of having eyes and ears with which to read the signs, with which to wrap ourselves within the great soul of the world.

Beyond Einstein: The Heart of the Matter

> The land is love. Love is what we fear.
> To disengage from the earth is our own oppression.
>   —*Terry Tempest Williams*

Archetypal patterns in the landscape are signs directed to the human soul saying "go this way." When we read the signs laid before our eyes, our direction, the trajectory and purpose of our soul on Earth—*in relationship with* Earth—becomes clarified. We begin to see where we are and where we are going.

What are the patterns that give rise to our lives? Are they the Golden Arches rising above the drive-through? We clearly read the signs of civilization, but what are they telling us? Are they codified by commercial interest? As if still tracking, we follow the signs, looking, questing, and creating the stories of our lives. Like our ancestors, we hunt for our lives, following the signs that say "go this way." But what way are we going?

I go to the eye doctor. I tell him that I am continuously writing, sitting in front of a computer and losing my clear sight. I need glasses. I tell him that I am writing about vision, and especially about the perception of relationships, thinking that he might get a good-hearted kick out of the irony of my situation. I imagine him responding by saying, "It's the stuff of the world that matters! Relationships, bah!" But without skipping a beat, he says, "Heck,

Einstein made a career out of it." I laugh, remembering Einstein's general theory of relativity. After some eighty years, I wonder why his theory is still illusive for most of us. Perhaps we have not fully grasped the nature of relativity because Einstein's insight was *the thought and theory* of relationship, and not the direct experience of it. Nonetheless, Einstein's theory initiated a change in the Western worldview. It shifted our perspective. We learned that we live in a continually changing and subjective world, that the whole world is relative and that only the speed of light is constant. This shook up our Newtonian and purely mechanistic worldview, but still, we have not fully grasped the look and feel of such fundamental relativity, the heart of the matter. Is this why we minimally relate and resonate with the Other, with other forms of wholeness in the world? Is this why we seem to be in such a mad search for a sense of wholeness?

I have already suggested the cultural conditions that predispose an attentional focus on objects, on man-made stuff. Our history of disembodiment makes a soul-satiating recognition of almost any relationship with the landscape nearly a joke. Furthermore, environmental degradation does not encourage a desire to connect, our rivers lined with broken bottles and plastic. All of this follows on something else, something deeper, something lodged more deeply in our psyches. Why is it that we still miss the implications of Einstein's tremendous insight? Why do we still, unconsciously and habitually, degrade the Others?

Perhaps the heart of the matter is fear.

Robert Greenway carefully defines relationship as "any exchange in which both or all participants are changed in at least
one way." By definition, being in relationship changes us. Do we
fear the change that is inevitably part of engaging in an honest relationship with the Others, the world, with the earth processes
upon which we are inescapably reliant? Are we fearful of confronting the fact of our own collusion in the destruction of our
planet, the ground of our being? If we look at the world with an
honest eye, will we have to step out of our comfortable cocoons in
order to be in relationship with the world as it truly is? At the core,
is our fear about being in an honest, deeper relationship? Is our
fear a fear of depth, the very dimension we most hunger for?

Chapter 5

Genuine Depth

Beyond Binocularity, Time, Space, and Fear

Depth, this mysterious dimension, . . . is the first, most primordial dimension, from which all others are abstracted. . . . The primordial experience of depth is always the experience of a sort of interiority of the external world, such that each thing I perceive seems to implicate everything else, so that things, landscapes, faces all have a coherence, all suggest a secret familiarity.

—David Abram

Deep Time (and Space)

> But all I could hear was the sea crashing against that cliff, half a
> billion years ago.
>
> —*Michael Collier*

I am looking into the Grand Canyon. A great opening in the earth
radiates red and buff colors and endless layers of rock carry my
eye into a chasm I cannot fathom. The scale of the depth, width,
and volume of the canyon throws me beyond the range of my typical sensations, my sense of balance. A subtle urge arises from
within, as if my body might irrationally throw itself into the vast
volume of space before me. I cannot stand so near to the edge. I
back away, speechless. My mind and perception are boggled.

I never fully understood the scientific explanation of depth perception. My professors in graduate school faithfully presented an
unconvincing and confusing mix of binocularity, retinal disparity,
and monocular depth cues as explanations. Retinal disparity describes the fact that—with our eyes separated by the bridge of the
nose—each retina receives a slightly different image of the world.
I was told that neurons in the visual cortex use this retinal information to "calculate" depth and distance. Monocular depth cues
were easier to understand. They were represented in every text-
book as railroad tracks converging in the distance, the gradations
of the blue color of distant mountains, and the superposition of
one object on the next. Overall, the perception of depth was presumably a combination of the two eyes converging, an unknown

number of monocular cues, and some rather magical computation within the visual system. But none of the mechanical descriptions of depth perception seemed to have anything to do with what happened to me on the edge of a canyon, or when looking into a star-studded night sky or even into a dappled forest. None of the explanations had anything to do with the response of my body to the spaces framed by the myriad things of the universe.

During the 1970s, David Hubel and Torsten Wiesel carefully mapped the distribution of neurons in the visual cortex as a function of their response properties. They found cells that responded exclusively to inputs from either the right or left eye, cells that were most responsive to one or the other eye, and binocular cells that responded to input from both eyes. Altogether, they characterized seven different degrees of response to left eye, right eye, or binocular inputs. The basically symmetrical distribution, however, was greatly affected by stitching one eye closed for periods of time. For cats and monkeys (and especially for kittens and young monkeys) whose visual experience was limited by the sutures—by monocular deprivation—the number of cells responding binocularly was quickly and greatly diminished or altogether obliterated. The implicit assumption was that these animals also had greatly diminished depth perception as a result of the deprivation of signals from the sutured eye. This was proof of the power of experience to determine the responsivity or function of neurons within the visual system—and presumably, in this case, to determine the perception of depth.

I had witnessed the ways in which perception changes as a function of experience while teaching vision improvement, and while picking at the ends of my hair at Namirama Girl's School in

Kenya. For this reason, an explanation based on experience made more sense to me than one based on retinal disparity or depth cues. With a slight conceptual stretch, the essence of Hubel and Wiesel's finding was that the quality of one's depth perception is highly responsive to incoming influence or experience. Like monkeys with one eye stitched closed, a habitual plane of focus diminishes the range of inputs to the visual cortex. A plane of focus at eighteen inches is a limited view of depth.

Jean Baudrillard, a contemporary French philosopher, calls the focal plane of our habitual perception our "surface of absorption." He suggests that it is a "superficial abyss of appearances" habitually viewed at the expense of "a hidden excess of meaning," and that it is oblivious to the seduction of the world.[1] From Ken Wilber's point of view, our common experience arises from what he calls "flatland," by which he means the ubiquitous forms of subtle reductionism: "Mind is reduced to brain, levels of quality are reduced to levels of quantity" and "depth is reduced to endless surfaces roaming a flat and faded system . . . in short, the multi-dimensional universe is rudely reduced to flatland."[2] Flatland is a perceptual world lacking an awareness of depth.

A superficial take on depth arises from superficial experience. We busy ourselves reading two-dimensional text; we travel on flat roads—where topography has been bulldozed out of our experience—past long lines of strip malls. The drive-through meal is now an American norm, convenience an American ideal, and money is the nearly singular measure of value. All of this minimizes our appreciation of depth. It is no surprise that we are "surface dwellers," as Brigitta, a sophisticated psychotherapist I know, suggests.

We have unconsciously stripped depth of its full meaning. For most of us a depth of field refers to a particular linear distance in front of us and to the plane of focus where the eye chooses to attend. Depth is commonly found eighteen inches from our noses, at the plane of our footfall, or, when driving, about thirty feet in front of us. But a depth of field surrounds us. It is more akin to lived experience, the body being *in depth,* than to a measured line. Visual perception is the dominant way we perceive the world. Certainly for humans, there is more cortical matter devoted to vision than to any of the other senses. When vision is limited by habitual planes of focus, so too is our common experience.

My thoughts meander like a river into the dualism that so pervades Western consciousness and perceived reality. No doubt, either/or consciousness translates into a form of flatland, devoid of texture or nuance.

I am wondering where I might find a "hidden excess of meaning" while walking in the Grand Canyon. My mind is cluttered with fragmented threads of Western history and dualism, with the ambiguous explanations of depth perception, and with a hunger for the look and feel of real depth. Tim, a young man who suddenly appears running down the trail, simplifies all of this for me. Tim has recently escaped from high school in New Jersey and knows all about the American suburban experience. I tell him that I am writing about depth perception and ask him about his experience in the canyon. He is clearly pleased to be asked and says that he has thought about this, especially since he and his friend have been backpacking in canyons for several months now. He tells me that in New Jersey, in the midst of gridded streets and rows of storefronts, he never saw more than two hundred yards. As he says this, he is looking into the vast canyon surrounding us, his eyes wandering over rock walls of every imaginable earth color, wrapping themselves around magnificent rock formations and looking into great caverns and crevices miles away. Here, in the Southwest, he says, he can see for a hundred miles. From the North Rim, he can see the San Francisco Peaks near Flagstaff. When his eyes rest on the distant peaks his whole body is able to rest. The effect "is like a muscle. When you hold it tight nothing can get through." He makes a fist, his bicep bulging, I must say, like a beautiful young buck. "But when you let go, well, then it flows." He adds that he really doesn't have words for the experience. "It just feels good."

Tim smiles as if he has discovered a major secret of the truly good life. He kneels on the sandbar where we stand next to Bright Angel Creek and makes a small sculpture out of stones some 270 millions years old, deep red or flaked and turquoise—the colors of time, a signature of depth. Tim doesn't realize it but he is exactly right. The muscles of the eye are literally at rest when looking into the distance. I am touched, watching him. I am also thinking about those of us who spend many hours looking at a two-dimensional computer screen eighteen inches from our eyes. I think of a colleague named Claudia. When I asked if her vision had changed since she began devoting her days to the Internet, she said, "Hmmm . . . I don't remember faces anymore." I doubt that she sees much depth either.

What am I getting at? Without experiencing depth in some fundamental way, as primary, how can we expect to be anything but superficial? We must experience depth to know depth. But then, what is depth? Is it a unique sensation, a particular impression on the body, a notion, or simply the shiftiness between near and far? Is it in front of us or in the mind? Or is the experience of depth to be found in the soul? Are we in it, or is it in us? In any case, depth is ambiguous. David Abram puts it this way: "Depth is always the dimension of ambiguity, confusion. . . . To the student of perception, the phenomenon of depth is the original ambiguity."[3]

Although Merleau-Ponty also refers to the ambiguity of depth, he carefully distinguishes between "objectified depth" and "primordial depth." Objectified depth is a measured distance and describes a location. It is detached from embodied experience. But as we shall see, genuine depth perception differs from distance precisely because it is beheld by the body. If depth is what exists between near and far, it is the sensing body that experiences and holds together the relation between here and there, this place and that horizon. And besides, the whole body feels depth, as if it is the space of *this* body, my body, responding and resonating with *that* space, as if a likeness is perceived, an affinity.

Merleau-Ponty's notion of primordial depth acknowledges this relationship. He describes primordial depth as "original" and indeterminant. It is the most existential of all dimensions, "like a medium, not like an axis or line," and irreducible to a third dimension—the z axis of Cartesian coordinates—that we traditionally associate with depth. Edward Casey, in a brilliant book entitled *Getting Back into Place,* says primordial depth "resists reduction not only to the spatial determinacy of a Cartesian triaxial analytical system but also to the temporal determinacy of clock time."[4] At this, my mind begins to spin. But Casey adds, "Such depth, then is therefore not a matter of space or time. It is a matter—perhaps even *the matter*—of place."

I wonder at the relationships between space and time, place and depth, experiences and events. These, I believe, are permutations on the same theme. Place is a space with experience added in; an event is analogous—it is time, a sequence of moments, with meaning added into the mix of moments. An event is the coalescence of myriad sensations and perceptions. It happens in a place and makes up what we call "a depth of experience." The dimension that arises from the gathering together of images, sounds, and scents is thus the essence of depth. Depth, then, becomes our lived experience, our experiences in both time and space, lived and made meaningful by the further deepening of events into experiences, the primary activity of soul.[5]

When depth is considered as eventful lived experience, and consequently not simply reducible to Cartesian coordinates described by height, width, and distance, the cues to depth become much expanded. They are no longer merely the convergence of railroad tracks, or the color of mountains along the horizon, but now include the horizon itself. The horizon, according to David Abram, is a structure within the visible world, within space, that communicates the abstract nature of time. It corresponds to that which is withheld, the future. It presents us with an opportunity to perceive an otherwise invisible future in an edge of landscape and thus to reconcile the concept of time and space as separable entities. The horizon embraces us in its surround and weaves this place and this moment with what is still withheld, yet to be discovered. The horizon holds together the near and the far, the present and the future. The horizon is thus a visible archetype, both holding and signifying a great depth of experience. As lived experience, in an embodied and attentive moment, the horizon, like a mantra, shows us that each step we take changes our relationship with the future. As a visible mantra, the horizon reminds us of the power of choice.

When depth is lived experience, depth cues also include parallax, a sensation that occurs when the perceiver moves. According to Gregory Bateson, parallax is "the appearance of movement in observed objects, which is created when the observer's eye moves relative to them."[6] But in addition to describing the perception of relative movement, parallax is a sensation within the whole body. It is seeing and absorbing the relationship between nearby willows passing by my body very quickly, the distant cottonwoods shifting through my sensory field more slowly, and the more distant red peak of rock passing by even more slowly. The embodied experience of parallax is volume. There is space, a full space, between these willows and those cottonwoods. By gazing with a little attention, the perceived volume is easily felt in the body, as if the spaces between the trees and rocks echo in the cavity of belly, inside and descending, moving as I move. The message is clear: I am in *it*—and *it* is both depth and flow. I am inside what Mihaly Csikszentmihalyi, author of *Flow: The Psychology of Optimal Experience*, calls the best of human experiences.

In the Grand Canyon, deep is a mile down and 2 billion years old. One billion, in my mind, is nearly conceivable. Donella Meadows, an environmental educator at Dartmouth College, describes 1 billion in terms of dollar bills: A four-inch stack of dollar bills equals $1,000. A stack 62.5 miles high equals 1 billion. One billion dollars, she says, would cover all the school, garbage, plowing, and road repair costs for her hometown (with a population of 2,500) for 750 years. She adds that the U.S. military spends 1 billion dollars in about thirty-two hours.[7]

But the Vishnu Schist at the west end of the Grand Canyon is twice this old. Between the upper edge of the canyon, the surface of the Kaibab Plateau, and the hard black Vishnu Schist are twenty-one layers of sedimentary rock telling a 2-billion-year story of the earth. At least 6 million years ago, 2.5 million years before what we call our human origins, the Grand Canyon was cut by the flow of water across the Kaibab Plateau, an uplift that arose some 45 million years before. What is now the Colorado River cut deeply through layers of earth—exposing the Kaibab and Toroweap Formations, the Cocinino Sandstone, the Hermit Formation and the Esplanade, the Supai Group, the Redwall and Muav Limestones, the Bright Angel Shale, and the Tapeats Sandstone. The passage of oceans and sandstorms—dunes piling up a thousand feet high—and volcanoes is all here, shifting my sense of this place, this land, and of the earth itself. The Tapeats Sandstone, the Muav Layer, and the Bright Angel Shale, for example, mark the presence and eastward movement of a shallow sea. The Tapeats Sandstone was the early edge of this sea, the place where quartz sand was laid into earth. But then the sea level rose, rivers from the east carried more sand and mud to shallow waters, and tides distributed sediments. The Bright Angel Shale is the mud, suspended in water longer than sand, carried farther from shore, and deposited among the teaming trilobites and worms in quiet, offshore waters. The Muav Layer reflects the continued eastward movement of the same shallow sea and the deposition of calcium carbonate, the stuff of seawater and seashells, from still farther offshore.

My mind tries to grasp the image of a sea lapping up against an Arizona landscape even before land plants existed, and the fact that at least seven such seas washed through the Grand Canyon.

Then I try to wrap my mind around the 2-billion-year story of Vishnu Schist, the "midnight rock" at the bottom of the canyon. My thoughts flounder—as if gasping for air. I cannot conceive the fact or implications of such depth but am left with much wonder and an inkling of what infinite might mean. Depth, it seems, is ultimately unfathomable. And yet, I might also say that depth is the fundamental perspective. In some sense, it is perspective itself.

Rigel, a star in the constellation Orion, sends its light to us from 1,000 light-years away. This is the distance light travels in 1,000 years, a quantity of time. In the abyss of the heavens, time and space are indistinguishable. We surrender our Cartesian formality in the interest of grasping, and entering into, a new dimension. Thinking of light-years allows us to expand our conception outward and into the universe. A light-year, we might note, is both time and space, a fundamental formulation of depth. Perhaps, then, it should come as no surprise that the entire night sky is filled with archetypes: Ursa Major and Ursa Minor, the bears; Cassiopeia, the queen; Draco, the dragon; Virgo, the virgin; Orion, the warrior; Cygnus, the swan; Aquilia, the eagle. The list continues and crosses every culture.

It seems that, universally, we look for and grant depth when a nearly visible dimension beckons to us, teasing our senses out. We reach to *make sense* of what we are perceiving, tugged by wonder. We recognize a form of dimensionality in the Other, the rock of many ages, the universe, the archetype—an image of our own soul. But the picture is beyond our immediate experience, both almost unseen and somehow revealed. Looking for depth is like reading a map, like looking *inside, over the edge,* and *into.* We might call this way of seeing "granting interiority." Is this genuine depth perception?

As if out of nowhere, or as if out of the very realm of depth I contemplate, I suddenly recall the ancestors. They come to mind, so it seems, to help me conceptualize the depth of the world within which I am embedded. Within indigenous cultures, the ancestral traditions are those that honor a depth of experience that is carried much beyond a single human lifetime. The ancestors thrive in the soil and rocks, watching, even "stalking" us with stories held by the land. But then I recall all the Others. I remember that the relationships between our hunter-gatherer ancestors and the animal Others are what made us who we are. My mind reels again with the recognition that my senses evolved in relation to the hunt, through learning to read the signs, the tracks. My mind is theirs, so old, so determined by and embedded in a depth of experience that I can hardly imagine. I can only bow my head in a moment of humility, hoping that I might be worthy of such a depth of experience woven through every fiber of my sensibility. As I ponder this, my sense of being is shifting and plummeting into a new depth of field.

Canyon Country

> In principle, extra "depth" in some metaphoric sense is to be expected whenever the information for the two descriptions is differently collected or differently coded.
>
> —*Gregory Bateson*

I'm leaving town in a hurry. I'm headed for canyon country, looking for depth. My body has been tied in a knot for a week and I can't seem to shake or stretch it into comfort. Like everybody I know, I have a cold. It comes and goes over many days. And even though I'm moving fast, I can't get out of town until midmorning—so much to do before breaking free, again like everybody I know. I leave fragments of messages and unanswered phone calls in my wake.

By midday, I'm two hours north. But I'm still racing around, breathless and doggedly doing last-minute errands in Flagstaff. I'm checking the oil and tire pressure. I'm buying flashlight batteries, another lighter, a pair of Patagonia socks on sale, then a pair of fleece sweats—all the stuff I think I must have in the backcountry. After nearly two hours of this, I feel the cold clutch of consumerism, as if she is incarnated and wrapping bony fingers around my mind while simultaneously squeezing my heart into contraction. I am desperate to escape her clutches. I tear out of town, headed north again and going so fast that I literally squeal and smoke to a stop at the last light leaving town. I end up in the middle of the intersection and everybody stares. My rearview mirror is filled with smoke, my nostrils with the stench of burning rubber. I nearly disappear below the dash. When the light finally turns green, I slink into a mechanic's shop to ask if I now need new tires. But mostly I need someone to talk to—I'm so jangled by my desperation and blindness. The mechanic tells me that I have just spent the half-lives of my nearly-new tires.

I'm in escape mode, on a quest and desperate for depth. I'm looking for a view with dimension, with scale and scope, and with the draw and density of any experience with real depth. What I'm looking for has got to be somewhere out there in the canyons, as far away as I can get from a culture gone superficial.

I wind around the west side of the San Francisco Peaks, travel west then north across the Navajo reservation, turn directly west to pass over the Colorado River at Marble Canyon, and travel

along the south side of the Vermilion Cliffs. The lower reaches of
the Kaibab Plateau stretch flat and wide to the south and suddenly
drop into a radical chasm, a sharp, dark ribbon that winds through
the plateau. I know the chasm to be the Grand Canyon, and I'm
headed for the North Rim. I climb the East Kaibab Monocline,
then drive south through deep ponderosa forests, wide, grassy
meadows, and golden aspens and spruce, all enshrined in late afternoon light. It is a labor of love, my effort to get there offered
and given back many times over.

I arrive at the North Rim Lodge—built of large stones during
the Roosevelt era, on the very edge of the canyon, and terribly romantic—at sunset, with not a moment to spare. I hurry out to the
west-facing terrace to watch the sun offer its last light. A dozen or
so people are taking photos. Most of them open the glass doors,
thrust themselves into the cool air, snap a flat photo, and duck
back inside without another glance. I wonder about snapshots and
the view from a camera. I remember our philosophic tradition, the
theories of visual scientists, and the temptation to liken the eye to
the camera obscura, a flat, two-dimensional surface—freed of at-
tachment, interpretation, or the coloration of imagination and
memory. The explosion in the popularity of photography, despite
the fact that it is one of the most highly toxic of industries, makes
me curious. It is as if we must have that photo, that canned com-
promise of beauty. Perhaps our need is for the reassurance that "we
were there," that we saw "it"—that beauty, that exotica, that travel
package that we have been conditioned to desire. It is as if we may
never witness such beauty again, our lives so filled with transience
and the 70-mile-per-hour fast lane. Thus deprived, we must have it,
as if beauty could be captured, held, framed on a flat surface, and
owned. Apparently, we have forgotten that we are co-creators

of whatever we choose to call beautiful, that beauty is "always already."

The next morning, I talk with Hannah, a robust and chatty tour guide on the North Rim. I ask her how people respond at the overlooks, where they are suddenly exposed to a wide view of a mile-deep chasm in the earth. Hannah smiles and says there are generally two responses: People become speechless or simply can't take it and back away. She says they more or less "change the subject." I am told that the average time spent on the South Rim of the canyon, where tourists arrive from all over the planet in ludicrous numbers, is twenty minutes. I understand this to mean that we simply don't get it, that we do not easily absorb the scale and scope, the implications of such depth; we leave for a hamburger— or any familiar food, comfort, or convenience. Is this no more than our minds mixing with an unfathomable reality, trying to make sense of what is beyond our sensory experience? How can we possibly fathom the depth displayed before us?

The first white explorers arrived at the edge of the Grand Canyon in 1540. They were a dispatch from Coronado's famous search for the legendary Seven Cities of Cíbola and their riches. Garcia Lopez de Cardeñas was sent to find a great river that the local Indians had described as being not far to the west. With Hopi guides, Cardeñas and his small band traveled for twenty days, emerging from a juniper woodland to view the canyon for the first time. Apparently they were more than staggered by the abyss. Despite the unprecedented 5,000-foot dive into the earth, revealing layers of rock and time previously unimaginable, they believed that they could reach the river in a day's time. They saw the river as no more than six feet wide, although the guides assured them that it was at least "half a league wide." The explorers managed to get less than a third of the way down to the river before returning to the rim and admitting that, in fact, the river was notably wide.

barrel into the canyon, descending through hundreds of millions of years, contemplating scale, deep time, and the depth of psyche, the sensations in my now infinitesimal body, and, ultimately, the nature of humility. As I reach the Bright Angel Shale, my favorite formation and destination, it is nearly time to turn around if I am going to make the trek back up before dark. But the Bright Angel Shale effervesces, teasing me with a turquoise shimmer, as if quietly vibrating with the lusciousness of a shallow sea, as if in remembrance of long-ago oceans. I can't stay: I have no flashlight, and the trail is often no more than two or three feet wide with sheer walls and sheer drops on either side. I have several thousand vertical feet to go and no more than three hours before dark.

I barrel back out, moving fast, breathing hard, and paradoxically falling further into the canyon as I ascend. I am hooked, as if strands of me are sweeping across the towering walls. Stripes of sedimentary rock emblazoned across the canyon walls stream by in surreal sensation. I see 2 billion years of history portrayed in magnificent earth colors and textures. I see several oceans washing over the land and then receding, leaving layers of sand and silt. I see the fossil traces of trilobites, the momentary movements of worms some 550 million years before. I see multiple edges rising and falling—in hundreds and thousands of radical feet—with every step. Nothing remains stationary as I move. The canyon, it seems, is alive. The vibrations of red and turquoise, salmon and tawny colors encircle me, reaching through open pores to my secret sensibilities, then pulsing there, echoing, reverberating, and restoring my memory of depth. I am full, satiated and reeling. I stagger a few times, gawking, and gawky. I am shaken by scale, my steps careful in a precipitous landscape where the weight of experience is everywhere vertical, up and down and *into*. My habitual defenses are now surrendered by the exertion and exhilaration of walking in depth, and my mind is now bewildered—as if something wild is seeping through.

That night, around a fire on the wide, east-facing terrace, I ask new acquaintances about the canyon and about the nature of depth. A handsome German artist, working in the competitive world of marketing, says, "Depth is an impression on the body." Yep. The wine dealer doesn't know what I'm talking about, and a couple from the Midwest admit that the canyon makes them nervous, somehow uneasy. They affirm my conjecture about why several people fall over the edge each year. I have guessed that on such an edge our senses are easily jangled, so unaccustomed are we to the sensations associated with depth on such a scale. As surface dwellers, we rarely see, experience, or even imagine such depth and we lose our balance. Most often, we simply retreat.

The next morning, I watch three ravens fly together just above the very edge of the canyon, with several thousand vertical feet as backdrop. One suddenly flips and flies upside down, jetting into a dive, then shifting direction at a right angle. This aerial display happens in seconds, and I can only assume that the brilliant black raven is either showing off or simply ecstatic. Gareth, a wilderness guide from Wales, walks slowly toward me across the stone terrace, as if stepping from the lodge into a surreal landscape. I ask him if the gambol oaks, so brilliant yellow now, mark the Toroweap Formation. We talk about his favorite canyon trees and the color yellow. Then, with a dreamy kind of look, he says, "Do you know the colors of ponderosa bark at sunset?" He moans a little, takes a breath, and says in his lilting Welsh accent, "They're colors to die for. . . ." I imagine the cottonwoods and box elders to the north, in a quiet canyon carved by a wash of water running through sandstone and gently slipping into the Escalante River. It is October and they will be many shades of yellow and green— "colors to die for," I know. Having been reminded of this by Gareth, I must go, now, this minute. I quickly pack up my breakfast, climb into the truck, and pull out the map.

The map I read is a mess of sweeps and tight repetitions of contour lines, depicting a topography of depth. Contour lines describe the shape of a land, depicting three dimensions on a flat plane of paper. But the map is never the territory. The map, with a little imagination, helps us to envision scales and dimensions beyond what the immediate reading of the landscape might reveal. Looking at my map of the Escalante, I see the territory to the north emerge and shimmer on the edge of my consciousness. The contours tease my mind's eye with great sweeps and bulges, cliffs carved in sandstone, and a ribbon of riparian green slipping through a canyon colored red, copper, and peach. I piece together the territory, being drawn, drawing the land in my mind's eye, making my map. And like the resolution of three dimensions from two, I wish to wring another dimension out of three. Three dimensions plus the sensations and sensibility of body and soul should satisfy my hunger. But I pause to wonder: Who draws our maps? Whose contours shape this land?

Sky Camp

There is a lot of unconscious resistance taking place, and a lot
of fear to overcome. One must travel to the other side of
fear . . . only then does true transformation really begin to
happen.

—*Malidoma Somé*

The acknowledgement and experience of fear is the door that
opens us to heightened presence and perception through which
we learn to live in the world as it is.

—*Laura Simms*

I am tripping into the Dreamtime.

The contours and roads, the many traces of our travels, are unfolding below me as I drive farther north, the plateaus rising and
falling, the rock spires on the edge of Bryce Canyon altering my
sense of reality. How could they be? What world is made of brilliant red and golden pinnacles rising like earth fingers and pointing to the gods?

Within moments after the sun drops below a nearby ridge, the
light of dusk becomes the cold of night. Within several more moments, I am buried in a sleeping bag. Soon I am wondering what I
am doing here, in a cold campground alone and wandering around
in Southwest canyon country. It is too early to sleep and my mind
wanders for several hours into the velvet black and infinite sky, my
thoughts rising and setting with the stars. The dark unknown is almost too visible for comfort, and too paradoxical with both den-

sity and light-years of space arcing above me. The mystery of it
pesters me and I cannot lie still. I finally unbury myself and trundle to the truck, dig through camping gear to a ratty old tape
recorder, then climb back into my bag with headphones and a
tape I have never listened to. It is of a conversation, taped nine or
ten months before, between me and Daniel, a serious astrologer.
Tonight, while gazing at stars and feeling lost in an abyss, he reminds me that I'm on a walkabout. Astrologically, my journey will
peak in three days. I had forgotten our conversation but give thanks
that, somehow, my stumbling path has put me on the right trail. I
imagine planetary forces aligning and compelling my plummet
into canyon country.

Who knows when a walkabout begins? This particular journey
began to take form in late August when Paul, the chairman of the
Environmental Studies Program at Prescott College, warned me
to watch out for mountain lions in the northern portion of the
Prescott National Forest. He knew I went there often and alone.
He had been doing research there all summer and said, matter-of-
factly, "You're deer size and they're hungry." Apparently, the
mountain lion population has skyrocketed, but for some reason,
there are few deer. He suggested I carry a gun.

I didn't take him seriously. Not until I began to see mountain
lion tracks as I entered the small canyon surrounding Harris Wash,
my destination. Then I remembered another conversation with
Justin, a tracker who had been a student in the course I had taught
in the same canyon a month earlier. When I told him I was returning alone, Justin said, "Watch out for the mountain lion." Other
than advising me to pee all around my camp to mark my boundaries, that was all he said.

I had spotted the perfect campsite during that trip. It was up

high, with a view of a large bend in the wash, and with the gift of
both early morning and late afternoon sunlight. To the west, the
campsite fell eighty feet straight down a red wall to a wide stretch
of sand and rich, riparian secrets. To the east, it fell equally as far
down a steep and sandy slope, dropping directly into Harris Wash,
thickly lined with cottonwoods. To the south, steep benches of
rounded slickrock receded and rose a thousand feet, and to the north,
a sandy trail led down to the wash through rabbitbrush and purple
asters. I would have a good view and could be barefoot all day.

   I had pined away for this sweet spot for a month, while doing
my academic duties at double speed so that I could return as soon
as possible. I was hopelessly drawn to this slip of water winding
through a small red canyon. But during the week before I left for
Harris Wash, images of Mountain Lion kept passing through me.
They were vivid and beyond my control. Mountain lions stared at
and stalked me. Sometimes I saw them pawing at me or batting me
around, and several times I saw lions land on my shoulders, grip-
ping my neck. I had experienced uncontrollable images running
through my mind before, but this was an invasion. I was unnerved
by the strength and persistence of the images. I uneasily wondered
what this was all about, told a few friends, and still left town as fast
as I could. There was no doubt in my mind—I was going to Harris Wash for medicine.

After a fast four-hour hike, I reach the sweet spot I have been
longing for. It is late afternoon and I make camp quickly as the sun
will soon set. As I put up the tent in the only possible place, I notice that there is a large stone, obviously carried up from the wash,
inches from the right edge of the tent door. On the tent side of the

stone I see scat. I look closely, remembering Juile telling me, just before leaving town, that mountain lion scat is full of hair and pinched, just like what I see before me. Out of respect for the lion's territory, or perhaps for superstitious reasons, I leave it untouched.

Then it is dark, daylight taken from the canyon like a sudden power failure. I am in the tent, hearing my breath and fiercely looking into flat blackness. I am frightened by the many mountain lion tracks I saw while hiking in today, by the echoes of Paul and Justin, the recent stories of women being attacked by cougars, and by the many images of Mountain Lion that invaded my mind's eye during the past two weeks, as the moon waxed. Now the moon is waning, rising nearly an hour later each night, and offering much less light. I pray for moonrise now. I pray for the light of Moon.

In the utter silence, I hear a big animal walking slowly through the sand around my camp. I keep staring into the flat darkness, trying to see by the meager light of stars. I am listening so hard that my stomach sounds like some small animal. I can hear my heart beat, the blood coursing through my ears. I hear my eyelashes touch the edge of my sleeping bag, and the Escalante River five or six miles away. And then suddenly, in the center of this silence, a loud and high-pitched wail cracks the night open—from both sides of the canyon. From the far side of the canyon the wailing continues, shattering the night. For several long minutes she calls, but at a fast run, the sound of her call quickly receding. And the one who was telepathically connected from my side of the canyon, from the edge of my camp, never calls again.

What am I to make of this? I have never heard such a sound, so eerie and piercing, slicing into me like a lightning bolt to the heart of my vulnerability. Jim Dale Vickery, author of *Wilderness Visionar-*

ies, says we must ache for vision, that we must grow cold, fall down scree, and be stormed upon. Scraping up a shred of cold comfort, I tell myself that I'm paying my dues. And then I remember the West African medicine man Malidoma Somé, insisting that a genuine initiation is frightening, as if one's life is in danger.

I have stepped into a symbolic landscape and the signs around me are immense. I am beginning to see languages rich with complexity and nuance. The winds speak as they pass through crisp cottonwood leaves, dry and brittle now. They rattle and call my attention, awaken me. I hear birdcalls announcing my arrival in the neighborhood. I hear wails in the night and find signatures of scat. When I get up late, after the moon has risen, to pee, I wonder how the lion will read this. What is cougar language? Are my pheromones pumping fear into the crisp night air? Does my scent reveal the depth of my vulnerability? What does she smell and hear and see? Who is she?

I make a medicine wheel in the morning. I had thought it would be in the sand below camp, large and expressive of ecstasy and gratitude. But no, I need medicine around and close to my tent, a sacred ring of protection. I choose four small and distinctive stones to mark the cardinal directions. As I move around the circle, I am just beginning to appreciate the alignments of the fundamental directions, the points of a compass. To the east, I see that a mound of rock, a pubis, and a thatch of cottonwoods, the fur, mark the downstream flow of water. A massive and graceful sandstone leg rises in the south, wrapping around to the southeast as if holding this curve of earth. Directly to the west a small opening in the top branches of two cottonwoods embraces a crescent moon, now setting over a lip of red rock. And turning to the north, I see a large, proud breast rising from the sheer vermilion wall. The wall forms a wide arc around the bend of sand and water below me. A great mother presides over all of this, a great "creatrix." She is wildly feminine and I identify with her. I am aligned, and in the center.

The day passes peacefully, as if the lion is out of the neighborhood.

On the third day, the day Daniel had intuited as the climax of my walkabout, I decide to hike the remaining five or six miles to the Escalante River. I wonder about this as I leave camp. Soon I am weaving through thick brush, wading through grass and fallen cottonwood leaves. I suddenly come upon the large body of a fallen cottonwood tree, arched and offering a doorway through the thick brush. Passing under it, I think, Ah, the threshold, knowing that I am stepping farther into a mythic landscape, walkingabout and wondering.

But the lion is in the neighborhood again. I feel watched as I splash through the wash and bushwhack my way across sandy terraces covered with willow and brilliant yellow cottonwoods. The sound of my wading, I imagine, is alerting her to my presence. She sees me. I walk fast with my head down, fearful. I wonder why my body is contracted, my mind projecting. Or is it that I am perceiving especially well, sensing a presence that is just below my visual threshold? I think of threshold experiments in visual science. As a subject I could not tell if I saw the light, intuited it, or made it up. I remember the shiftiness of my threshold and the odd feeling that arose with the recognition of it. Even then I wondered about the edge of reality. Where is it? And why do we think that vision is the final word, that only seeing is believing? I don't see the lion, but I know she is here. I feel her seeing me. I walk miles to the river through a quiet flow of water, contemplating the mysterious depth I'm in and knowing that she knows exactly where I am.

But I can no longer stand the fear I feel. "Damn it," I say to myself, "look!" I make myself look between the feathery stands of willow, thickets of tamarisk, and branches of Russian olive, believing that I will see her watching me. I look, bravely penetrating the shadows, looking *between* and *into*. My eyes wind through willows, drawn by what is there and not there, in the spaces between. The spaces are filled with velvet-textured light, a collage of deep green and dark earth colors, splashes of yellow and sometimes a spot of blue, or deep red from sandstone curving in and out of the distance. My eyes love the look and feel of this and soon I am seduced by the whisper and bend of willows, the spread of rock, the secret places tucked into thickets. I am enveloped and enchanted—and emboldened by the power of my eyes to see.

I walk this way, in beauty, for several more miles, returning to camp. Perhaps a mile from camp—just before passing back through the cottonwood threshold—I hear myself asking for peace with the lion. A comforting answer comes quickly and I walk into the bend that marks my camp at ease. I make a simple meal, watch the stars appear in an endless dark void of sky, and sleep well—oh-so-deeply, without fear.

When we attend to the spaces between the things of the world, the backdrop—the things and places that we have minimally no-

ticed—moves forward into the field of our vision. The backdrop
becomes our primary awareness. The shape and substance of the
land, the red rock wall or space of grass, fully enter into our awareness and become the visible and salient. Our attention widens naturally and our eyes now penetrate more deeply into the world.
Looking into the branches and trunks, between the things of the
world, is a fundamental practice for opening our attentional focus,
for stretching ourselves beyond our predominant view, our habits,
our fearful projections. Opening like this, we experience a kind of
spaciousness. We see a new coherence, a new wholeness. We see
the fall of land, the boulders behind, a new story. My friend Hilde-
garde has much to say about this. A year before, I had asked her
what she had learned about seeing. We were strolling through an
exhibition of contemporary Northwest painters. She is a painter
herself, with seventy-some years of attentive looking and seeing.
She smiled and said with much reassurance, and as if she were talking about divinity, "Find the negative space." She tells me that
when we look to the places between, the negative space becomes
what we see, the figure in the ground. She says, "Look between the
legs. It reverses the world."

In the morning, I go for water, walking barefoot down the trail to
the wash. Fresh scat, like nothing I have ever seen before—velvet
black and shiny, puddled and placed an inch from where I peed
the night before—sits at the bottom of the path leading to my
camp. My first thought is that it is time for me to leave. In the language of scat, somebody is feeling squeezed. I ponder this, and the
gathering rain clouds, while I fill water bottles. It is time to leave.
Returning to camp, I circle the medicine wheel one more time,

praying to the four directions. First I bow to the east. As I raise my eyes, I am humbled by not knowing the name of the first plant I gaze upon. I pray that I will always be willing to begin anew, like the sun rising each day. I contemplate beginner's mind, looking further into the east, the place of new beginnings. Below me, the wash flows eastward, down, down, slowly and gracefully through great curves in red and flesh-colored rock. I pray that I may always know and be part of the flow. Again, I pray that I will always be willing to begin anew. I contemplate renewal and flow as one, and am grateful. I bow again.

To the south, red sandstone rises to a soft peak, as if a knee rests on the southern horizon. A band of sandstone descends to my eye—a wide thigh spreading itself before me. I see, again, that I am riding on a great leg of this earth, abundant and strong. I imagine the strength of my own legs and envision the ancient Mayan messengers, running with news. I pray that I will be graceful, my legs stretched and strong. I pray that I will carry the message that so compels me with strong legs, that I am worthy of walking this sacred ground.

I bow to the west. As my gaze rises, I again see the notch created by the top branches between two cottonwoods. Blue sky streams through the opening. I pray that I will see the opportunities, the openings, and be willing to step through them, to be fearless, even as the moon now sets, or as the sun sets, as it grows dark. I pray that I will be fearless and willing. A wave passes through me as I recognize the depth of my fear—and the warrioress that I am. I bow in gratitude for the recognition.

To the north, the prominent red rock breast presides over this wide curve of rock and sand and water. I pray that I will be as vast and proud as this solid red rock, to be as solid and giving, continually surrendering to water and wind, eroding, and giving myself away. I pray to be both solid and giving. To be yielding and tough, as hard as rock.

I look to sky and suddenly feel my feet now solidly on the ground.

I circle again, and again, only to give thanks. My knees tremble and I cry with the power of prayer. I am in the center now, aligned with planetary forces, and weeping with humility and gratitude, and because this is real life. I sit. But the voice, now loud and insistent, says, "Walk out now." And then again, "Walk out, now!" I do, despairing in the leave-taking.

I arrive at the trailhead in the late afternoon. I watch myself circle the truck, stalking it with a bit of wildness and suspicion, as if it were a danger. I put a single boot in the back then walk away, change my shirt, then walk away again, back toward the sunny wash. My arrival is really an approach, and something akin to passive and aggressive. I am somehow confused, my body pulled by the golden light spilling over sand and resistant to the fact of a truck. But I get in and slowly drive down the thirty miles to pavement. Although it is late, I am not in a hurry.

Eventually, I make it to the campground I imagined as the next stop. It is full of campers and cement. I cannot stay despite the attendant's kind offer to camp in the hard-packed dirt of the "overflow area." In spite of the growing darkness, I drive away into miles of slickrock. As I pass a dirt road leading into what looks like the very depths of a dark forest, I catch a glimpse of a homemade sign saying "Bed and Breakfast: Bill and Sioux Cochran, proprietors." The voice I am beginning to simply call "the walkabout" tells me to turn left, driving into the Dixie National Forest at dusk. I end up at a funky, log ranch house many miles down a dirt road. The couple running the place are a riot. Sioux tells me they are "ranching in the nineties" by feeding and bedding people. She feeds me turkey and mashed potatoes, gravy and biscuits. Bill tells me the local ranchers in the area have killed off the big male mountain lions. Adult males, he says, keep the younger males from mating, but now that they are gone, the young males are mating like crazy. The population has skyrocketed. With so many mountain lions, the deer are gone. He says the ranchers should keep shooting the lions. "They're out of control." He worries that one of his three sons could be taken. "They're hungry," he says.

I wonder about the edge between what is real and what is unreal, or between material reality and projected reality, or between the seen and the unseen—and the power of our beliefs to shift that edge. I wonder at the way we tend to confuse the unseen with the unreal. These are not necessarily equivalents. The unseen lion was real in my imagination, possibly real in the physical world. The reality was that I acted as if she were real. The reality was her power to make me tremble and shift my view of the world, to see sunlight and secrets between branches and brush, the beauty of the world only slightly veiled by my fear. This shift in the shape of things matters most. Is this, then, the most real? A week later, minutes after I return home, the phone rings. It is Gareth, the wilderness guide I met in the Grand Canyon. To my surprise, I find myself describing the shiny black scat left on the footpath to my camp. He says, not knowing anything about my story, "That is mountain lion scat, dropped right after eating the heart, lungs, and liver of a recent kill." At this point, the mountain lion is real enough.

These notions of reality are sliding scales, thresholds that make

our relationship with the land what it is. Our beliefs mark the edge
between inner and outer forms of reality. At this edge, we amplify
one reality over another. It is a permeable edge, a zone of gradations between black-and-white reality where our imaginations fill
in the unnamed spaces. Do we fill the space with images born of
Little Red Riding Hood being teased and tricked by a nasty wolf
in the woods, or do we see Changing Woman, receiving Sun as
her lover? Do we imagine changing as we open ourselves to the
potent power of sunlight?

After leaving Sioux and Bill's bed and breakfast in the Dixie National Forest, I meet David at the desk of a small inn in a small
town. The town is noted for being the last in the nation to receive
mail service, in 1940. Although David appears to be a worldly and
hip young man, he has gone rural—called by the canyons surrounding this small town of settlers. Being in canyons, he says, is
"getting closer to the pulse." He looks at me with serious anticipa-
tion, waiting for a response. I nod and from behind the computer
where he works, he mumbles something about "the mother." Then
he says, "feminine," and begins to talk about being *inside,* and that
it's frightening. "Some people can't take it," he says. I think about
the mountain lion and assume that he's talking about the feeling of
being watched from all sides, from high walls and from behind
thick stands of willow. Or perhaps he's referring to the fact that
within steep canyons, the most provocative place of all, horizons
and edges rise and fall with every step, with every new glance. We
become edge-y and lose our sure footing. Then, I think, he is talking about the shadows so apparent in canyon country, and the way
light leaves a canyon quickly, as if the world is suddenly plunging

us into darkness, as if the light has been switched off, the lover's eyes hidden.

But then I hear him. He is referring to seduction—to his own, his radical move to canyon country. He's talking about a tug on the senses, to being tugged out of himself. Yep, we lose ourselves, our old sense of self replaced by an infusion of the world, the senses so wildly renewed. And for some, this too is frightening.

Eros, Earth, and Sky

> If human consciousness can be rejoined not only with the human body but with the body of earth, what seems incipient in the reunion is the recovery of meaning within existence that will infuse every kind of meeting between self and the universe, even in the most daily acts, with an Eros, a palpable love, that is also sacred.
>
> —*Susan Griffin*

In canyon country, we are both inside and penetrated by the powers of earth. Being inside, we are held within territory beyond, or bigger than, ourselves. Our step is sometimes uncertain, our senses alert and awakened. We are called out—pouring emanations of attention and self into the landscape—and sometimes we "lose" ourselves. This might be frightening—and it might be ecstatic, erotic.

Light is electromagnetic energy, one of the primary penetrating forces. Once cast from the sun, light streams onto planet Earth like liquid. It freely enters us as both photon and wave. In daylight, we ingest billions of photons during every open-eyed moment. As waves, light vibrates into us. It is transduced, within several layers of retinal tissue, into electrobiochemical pulses surging deeply into the tissue within the mammalian brain. It shifts from wavelength to color, from photons to pulse—and from pulses to hormonal flushes. In my mind, it's a sexy event—one that is most fully felt in springtime, when light-filled days grow longer and animals mate. We call it spring fever. But it is light coursing down a direct neural line from the retina to the pineal gland. We call the pineal gland "the master gland," not realizing that its reception of light has everything to do with the unfettered creative energy we feel in springtime. When stimulated by light, the pineal gland releases a cascade of hormones, drenching the body in hunger, thirst, or great desire. In this way, the pineal is truly an interface between the visible world, the mind, and the body. The third eye of many cultures, in the center of the forehead, points to the place of the pineal gland. Buried in the brain behind the third eye, it is the most highly protected organ in the body—perhaps the most essential, or the most intimate. For some it is the very seat of the soul, the "eye" that informs us most profoundly, its signals spilling through the body.

Light reflects off every surface, surrounding and drenching us with stimulation. We are soaked with light of many vibrations— blue light, red light, a multitude of greens, the purity of yellow. These are perceptual worlds created by particular energies, by variations in wavelength. The variations correspond to the speed or frequency at which each form of energy moves and enters us. Ultraviolet light, closest to the wavelength corresponding to blue, literally fragments and unweaves the integrity of the retina with the intensity of its pulse. Red light moves much more slowly; we feel its vibration as heat. Even hints of peach and rose colors are felt as warm. Landscapes drenching us in warm colors—flesh- and salmon-colored boulder fields—are unmistakably erotic, as if low-frequency light is made for the root chakra, resonating with a slow vibration found deep within the body.

Each color corresponds to its own wavelength, its own vibration reaching into the body. Our bodies respond as if in silent and continuous conversation. In the receptive body, light descends, traveling through the chakras, each chakra resonating with a particular vibration. Green in a dense and saturated forest pounds on the door to the heart chakra. Yellow light travels down and swirls through the solar plexus, the third chakra. And red light, a slow vibration, further descends to the root, dense now with desire. Who can say that this gathering of vibrations is not deep and naturally erotic? We receive, our light receptors as metabolically demanding as any structure in the body, our hearts hungry for beauty and resonance.

Light carries signals from earth and sky to eye and heart—the heart pounds in response, the lungs inhale, inspiring in the clear light of beauty. Light from earth and sky surges into the pineal gland like a seasonal tide—a flood of hormones touching the core and fingering the root chakra. In the end, we are light bodies, so deeply touched and stimulated by light running through the body. But is it not Eros that fingers the instruments that we are? Like light, Eros touches our core, descending and then arising again. Like erosion, like the flow of water running downhill, burrowing into the earth and creating chasms and contours, the canyons and secret places, like this, Eros and light burrow into my core. And like mountains that arise with magma and the deep heave of the earth, that arise with the downward flow of river and creek, eroding tonnages of sand and allowing the keel of a mountain to float higher on the mantle of Earth—like this, Eros and light pulse together, one making the other. Like water following the thrust of mountains, the shifting and folding of tectonic plates into monoclines and anteclines, turning lines of sedimentary rock into ripples and sweeps that carry my sensitive eye to sky, or to diving below, deeply into earth—like this, light and Eros are one movement, the pulse of earth.

This wide pulse, like the swaying hips of a great mother, is visible in million-year increments, in oceans flooding and receding seven times over the Grand Canyon. I see tectonic plates bumping and grinding, volcanoes spewing themselves, erupting and offering the earth's core, the very hottest. We too are volcanic—yet our cores, the great heat of *our* bodies and souls, often remain unseen and untouched.

In *The Erotic Impulse,* Mary Klein claims that we are erotically crippled. She calls our collective fear of eroticism the cruelest abuse of all. The price we pay for our fear and denial, the loss we live, is the loss of pleasure, the very ground of love. Without an erotic sensibility, we suffer the agonies of sexual and relational dysfunction, of relational disjunction. Our relationships are diminished, limited by the body of the earth—what we rather coldly call "the environment"—being so far away from our touch, our senses disengaged. Sensuality becomes merely sexuality and given two-dimensional, dualistic consciousness, we are either turned on or off, good or bad.

But eroticism is an *entire realm* of relational experience. For the

Australian Aborigines, the erotic impulse is born of wind passing
gently through a mound of pubic hair, or lovers communicating
telepathically in the invisible realm. For the Dineh, Changing
Woman—spreading her legs to receive the light of Sun, becoming
impregnated and learning the way of relationship—is erotic. For
Joanna Frueh, an art historian and performance artist, the erotic is
"socially transformative."[8]

The erotic powers arise from down deep and become manifest
in the twining of light and sinew and soul. Yet having forgotten
the body, having forsaken a depth of experience, we wither, our
souls unsatisfied. But the remembrance of depth is *there*, its possibility and existence offered by looking *there*, between the things of
the world. We *see* depth. A depth of field is *there*, almost hidden
from the modern mind—but look, what you see, *there*, reverberates
in the body, inside. Like a synaesthetic wave, we feel the depth we
see within the body.

The erotic is a realm born of the sensate energies that tie and
twine bodies into one. If we wish to participate, we must open ourselves to the vibrations within which we are embedded, and give
back. For Deena Metzger, the erotic powers are receptivity, resonance, and radiance. In my experience, these are pragmatic powers. Receptivity is not simply a good idea. It is something we *must*
do. Receptivity makes possible the resonance between ourselves
and the world, and thus reveals our true alignments.

Aligned and centered, we know our place. Aligned with beauty,
I open in a kind of whole-body gasp. Encircled by beauty, my
senses cannot help but participate, reaching out for more, grasping and despairing as I leave canyon country. People argue about
beauty. Is it in the eye of the beholder or is it a property of the be-

holden? From a relational perspective, the question is clearly dualistic, cast in either/or terms. If there is an answer, it is clearly *both*. Beauty is in the thing seen, *and* we grant beauty, making that with which we identify, in some resonant way, beautiful. Beauty is the experience of a shared vibration resonating between the thing seen and the one who sees—between color, form, and texture and the many fibers running through head, heart, and body. Beauty, then, might best be named in terms of resonance, a vibration that moves us, that makes us see and do differently. We bow, cry, or give thanks. We want more and soften in order to get it. We participate. Beauty becomes a call for engagement.

"Beauty conveys us," says my friend Barbie, a prominent lawyer crusading for justice. She means that true justice is beautiful and that, when justice is served, it transports us, carrying us outside of ourselves into a participatory realm. We become more deeply embedded selves, subject to and gifted by being in a voluminous and animated depth of field. Under the spell of beauty, we too vibrate, radiating like light itself. We recognize this feeling, our *own* radiance, as an erotic power, and we know our alignments. We are no longer alone but emboldened by being inside and part of Spider Woman's wide web of being. Depth perception, at this point, is alive and well.

This is the function of beauty, reaching well beyond the question of where beauty resides. There are many ways to say this. James Hillman says, "Beauty is an epistemological necessity; it is the way in which the gods touch our senses, reach the heart and attract us into life."[9] Or, like Robinson Jeffers, I might say that we "fall in love outwards," being seduced by big beauty and ecstatic with recognition and kinship. We fall in love, wanting more of this. We look for it, and radiating, we create it. We are then walking a path of beauty and the world we see and touch and taste is transformed.

Light is streaming into my eyes. It is entering through every possible pore. It is burnished yellow, and a brilliant, pure yellow. It is golden and ocher, harmonizing and teasing my eye into a greater depth of field. My eyes follow the many shades. They are bright yellow, then perfect yellow, unique in all the world. Then I see yellow flavored with lime. Next I catch the color of oak leaves in mid-October, still green but now touched by yellow. I see rich, dense green, the color of grass, juniper green, deep pine green, and pure green. Dark sage, pale sage, and white sage appear. My eye travels and with every small adjustment from one golden or green leaf to the next, my sense of depth becomes ever more finely tuned—more carefully calibrated. Multiple planes of focus now beckon me, seducing me into an awakening. I am called out and hungering for more, filled with light and gratitude, and I am changing.

Chapter 6

Visionary Practice

Reversing the World

Use the vision farther than you can see.
If you put your mind with your eyes,
you can see a very long way, and your heart, too.

—Bobby Billie, Seminole traditionalist

Local Visionaries

Put things in order before they exist.

—*Tao Te Ching*

I hear traces of visionary consciousness in conversations all around me. The hints I hear have led me, as if sweetly by the hand, to the conclusion that the visionary resides in many of us, perhaps in all of us who are perceptive and looking toward an illuminated future. My student Scott, for instance, says, "An intimate look at what is immediately in front of us becomes the vision of infinite stories which are still to be told." Michelle, an engineering student who left mathematics for the speculative landscapes of ecopsychology, says, "The visionary must be dissipative." She is referring to the notion of dissipative structures described in systems theory, placing the visionary within an energetic feedback loop with the light, sound, and scents that enter her. She says, "The energy coming through her senses becomes apparent through creative and imaginative gestures." Dmitra is another student of mine. I hear her say, "I'm compelled to sustain the colors vibrating in the mud beneath my feet, rising up my spine, licking my hypothalamus, and shooting out my head. This is how I care." She is young and looks into the world with a romantic intensity, always asking and always risking. When she asks for guidance, I tell her to spend a weekend alone in the Prescott National Forest. She comes back brilliant, shining like light. I look at her radiance and see a local visionary in the making.

In 1995, the *Utne Reader* chose one hundred visionaries for the "liveliness, richness, conviction, and clarity of their vision." The *Utne Reader* described these individuals as expressing a "contemporary creativity. . . . They think about our crises, collective and personal; they ponder values and pleasures and hopes; and they offer vivid, memorable answers to questions that concern us all." Visionaries do not attribute relevant answers to a single person or genius; they are "positively allergic to the idea of the hero-thinker" and they "go out of their way to credit collaborative effort, obscure colleagues, and family, friends, and other everyday people with great, even decisive powers of inspiration." Furthermore, they "relish paradox" and "show a remarkable willingness to go beyond the boundaries of their personal agendas."[1]

Ultimately, the visionary is one who sees well—with clarity, precision, and imagination. She sees the world as it is, freed of her projection or personal agenda, and yet also illuminated and embellished with the cultivated capacity of her imagination. She sees *into*, her vision flexible, responsive, and roving across inner and outer landscapes and dimensions, stretching between visible and invisible realms with the potent power of imagination. She is engaged with the world, on the lookout for signs, for patterns pointing toward the Beauty Way. Like a sushi chef with an eye trained toward aesthetic detail, the visionary is tuned to the aesthetics of any circumstance, to the way in which color and form "pattern up" and make wholeness. In a patterned world, new forms and properties emerge and she sees what is possible within her immediate view. She attends to the edges, where things change from one thing to the next and consciousness expands, where the world becomes more than it initially appears to be. In this way, the visionary pulls the great field of potential into greater visibility. She thus re-visions and reconstitutes the world with perceptual practice and prowess.

I am piecing together a profile of the contemporary visionary. I hear all sorts of hints in the nuances of conversation. Neal Mangham, the president of Prescott College, my professional home, says, "The industrialized mind has lost the art of being still." The price we pay, says Mangham, is that "the world no longer informs us. There is little opportunity for the patterns to emerge," allowing us to "build whole new truths out of apparently unrelated pieces." I hear my friend Liz Faller say, "Our contemporary problems require that we see beyond our conclusions." Liz is a dancer and believes that we must tune ourselves to the "frequencies and vibrations," the light and the rhythms. My friend Bill Plotkin, a seasoned therapist and Vision Quest guide for some twenty years, says that we must learn to "see into the dark," to illuminate the shadows, to dig deeper into our souls.

After hearing such juicy comments surface in casual conversation, I began asking outright: "Who is the contemporary visionary, anyway?" Without skipping a beat, Mangham tells me that the visionary gives voice to what is yet unspoken and unseen, "the new truths." As the president of a progressive college, Mangham is often asked to state his vision of education. Mangham says that this question is impossible to answer because any true vision must keep changing, continuing to shift and emerge. His vision for Prescott College is clearly a process, not a static picture in his mind. He describes this process as "finding the heart of the institution and giving it voice." The process is a matter of discovering, or *uncovering*, an image—not *having* one. Then, in his own understated fashion, Mangham adds that his grandfather taught him, while hunting, that the world is to be discovered in one's peripheral vision. The image to be revealed, says Mangham—the right metaphor, the heart of the matter—arises by looking off to the side, to the edge of one's vision, where everything changes.

Plotkin lists three qualities of the visionary. First of all, he says, the visionary must be able to access the images that are generated by the soul, the images corresponding to the heart's greatest desire. Second, "Visionaries must have a compassionate understanding of the world. They must recognize the dynamic and interrelated nature of reality; they must perceive the web-work." The third quality of the visionary is the ability to "pattern up" the images that arise from the soul with those in the world. The visionary, Plotkin says, has a "creative and synthetic ability to see how personal images can be effectively embodied in the world. She has a keen understanding of both her unique gift and the world's need, and she can creatively bridge the gap between the two. This is the path of greatest service." He then describes a particular and metaphoric way of seeing—the way that "patterns are like one another." He says it is like crossing disciplinary boundaries and finding that one's set of metaphors applies in interesting ways across the boundaries, as the patterns both shift and match. Doing this is like skipping across realms of experience and distinctly seeing the patterns that connect. As an example, he mentions the way principles in physics have been brought to psychology: We say "quantum leap" to describe radical shifts in one's level of understanding and we refer to the "holographic self."

Fractals provide another example of crossing disciplinary boundaries. Fractals, a concept from chaos theory, are self-similar patterns across scale changes. In my mind, Plotkin is describing "fractal consciousness," a form of perceptual flexibility in which the mind is loosened from dualistic ways of perceiving. In this state of mind one more readily perceives similar patterns across physical scales and psychological dimensions. Fractal consciousness allows one to see visual contrast effects—for example, in relation to contrasting ecological zones, testy national borders, "culture shock," and "being on an edge." We might say that these phenomena are all "edge effects." Like Plotkin, I consider fractal consciousness an essential capacity for visionary viewing. It requires an open imagination, making peace with a kind of "loosened ambiguity"[2] in the mind. It is a poetic mode of consciousness in which one is able to make perceptual leaps—as if casting one's eye through a doorway, as if looking for forms of connection or alignment.

Plotkin adds that the visionary honors both light and dark experience. Light experience is that of the day world, the world filled with light and all that is seen. In Plotkin's conception of the medicine wheel or, as it is more universally known, the "wheel of life,"[3] the dark is associated with the West, the place of sunset and twilight, the moment of going into the darkness of night. It is the place of the visionary, of inward reflection, of the soul, and "all ways of not-knowing." The visionary, he says, "has to be willing to venture into the unknown—like Psyche looking for the love of Eros—and to bring back into the day-world what is yet unknown. The visionary illuminates the unconscious."

The imagination, says Plotkin, is what allows us to look into the unknown, into the dark. A "refined imaginal capacity is essential" to seeing what is yet possible. "The old rules and algorithms aren't going to help for the evolving work of the visionary. New patterns and possibilities need to be brought forth." Plotkin adds that the visionary has the courage to be an activist, to agitate for cultural transformation. However, "she is not a rebel who merely reacts. She is a non-conformist." She has chosen between her own vision and that which is conditioned. "Her vision takes precedence over tradition, over the prescribed form and pattern."

I ask my friends David and Delisa: "Who is the visionary?" Delisa is a dancer. David is her partner and a genius mechanic, working on the fastest racing cars in the country whenever he needs to make several hundred dollars in an hour. Otherwise, he gardens, sculpts, and designs water and glass prisms at home. I think of him as a wizard.

The three of us stand in the golden glow of late afternoon, so engaged in conversation that our walk up a sandy wash has come to a standstill. Delisa is drawing circles and designs in the sand with a stick. She lights up when I mention imagination as an essential quality of the visionary. She says that it is only with the imagination that we are able to follow energy through the body, and that this is how we begin to feel the inside of ourselves once again. Delisa teaches modern dance at Prescott College. Her students learn to "tune into both the inside and the outside at once" as if to a single stream. From here, they learn to move together as a group, as if "flocking." Delisa says, "Flocking is a way of sensitizing individuals to lightning-quick responses, to feel the group as a connected organism. It develops a sensitivity of the entire body— feeling with the back side of the body, feeling the flow of electricity and communication between bodies." Delisa describes how she does this with students: "I have them gather in a cluster and give them vocal commands: right, left, turn. They are to stay in a tight group but not touch each other. Then we repeat the same exercise but the directions are called out by members of the group. Then we do the exercise without vocal cues, responding only to the group flow, and then with our eyes closed." The students become increasingly attentive and sensitive to one another as they go through this progression. Delisa is glowing now, radiating with the mere thought of dancing.

David gracefully sidesteps my suggestion that he is the kind of visionary that I am looking for but does admit that he experiences himself as a "continuum of awareness" and as "a functionary of a larger being." He says that he is here to create, as a "materializing element." I ask him how he might recognize the visionary in others. He says, "Look to see who is doing that which, *by God*, they want to do." He laughs in the light and emphasizes that the doing is by God. As the sun bathes us, I realize why I could not interview David and Delisa apart from one another; on some level, they are visionaries for each other. There is no doubt that she is a dancer directed by God—she is enlightened when she speaks of her practice, of the dance—and he is thoroughly tuned to "both the inside and outside at once."

Don is another local visionary, one who lives on several hundred acres of open grassland and juniper-piñon woodland with his family. He carves twelve-foot totem poles, sculpts, paints, and builds houses, sheds, and barns out of mud and cement, all shaped into great curves that rise and dive and frame mountains and sky and grasses.

Don says that the imagination allows us to see various "totalities." He builds everything in his mind, whatever he wishes, before beginning any project. "Fundamentalism lacks vision," he says, "because the Bible always has the last word. Any real vision must be generative and forward looking." The visionary sees what is possible; he "sees invisible things as reality." He praises Rudolph Steiner, the author and visionary who inspired Waldorf education, for his ability to see spiritual things physically, to see a person's aura like a radiant smile or frown, a distinct view of one's soul. As a child, Steiner apparently thought that everyone had the ability to see such things. "But it takes practice," says Don, "beginning with knowing the power of one's imagination."

Good practice for the visionary, says Don, is to go outside on a dark night. He chuckles and talks a bit about demons and the creative power of our imaginations, concluding that sight can be an impediment to visioning. In darkness, one's imagination rises to the surface in full flower, full of the mythic, full of hope, and full of fear. In darkness, we see the many faces of our imaginations.

Don laughs again and says that the visionary sees clouds before they arrive. He senses them gathering overhead—in the near future—by the quick and subtle changes in temperature. Then he turns to his wife, Becky, and nonchalantly says, "There's a cold front coming." He is reading the signs, the nuances in the evening air, the feel in his bones. He is both aligning with how the world really is and imaging a future state.

Becky then tells me the story of what they call the Bootlegger hike. A band of kids and several adults were celebrating a birthday by wandering around in the national forest that borders their property. The birthday game was to find the way home while blindfolded. Becky's sons, Kanin and Cody, who were ten and twelve at the time, walked home as cool as could be, blowing the minds of everyone who witnessed.

Kanin and Cody, who are sitting at the dinner table as Becky is telling the story, shy away, laughing at how silly the whole thing is. It was no big deal to them; they simply followed the sun's warmth, the wind's direction, and the flow of the streams they crossed. Now aged thirteen and fifteen, they wonder why this story should be in a book on vision. "Being blindfolded is good visionary practice," I say. They either pretend innocence or simply do not realize that the visionary must be sensitive to nuance and must learn to see in the dark.

The visionary is progressive in her thought, thinking into the future and visualizing possibilities. Contemporary visionaries move with Others, recognizing a relational world and "flocking." It "takes practice," say Delisa and Don. And as David made clear, the visionary is directed by God and dancing, doing what she most wants to be doing; the heart is behind the whole act. As Mangham suggested, the visionary gives voice to the still buried heart, hearing the hidden, inherent desires for wholeness in others. She perceives what is yet unseen while looking into the world, while sensing a snap of coolness in the air, like Don. She sees that which is possible embedded in what is real, bridging between seen and unseen realms with memory and imagination. And like Psyche, Persephone, and Demeter, visionaries descend to the underworld, going "into the dark" on revelatory missions.

I dream of visionaries living in every neighborhood. In these uncertain and potentially disastrous times, the power of creative vision is crucial.

Imagine This

> Let the world of rationalization and of the senses be consumed
> in the fires of imagination. Free the eternal soul; let it taste again
> Infinity.
>
> —*William Blake*

A near quarter moon hovers in the late afternoon sky. My imagination wraps around the back of that lovely, luminous body. It is
spherical and dark on the back side. I visualize the entire moon
and know that only one-quarter of it is lit, my imagination revealing to me what is most real. Imagination, it follows, is not necessarily unreal—as is often assumed. It is not right or wrong, real or
not real. Rather, imagination is a mode of consciousness, a unique
capacity of the mind and the "deepest voice of the soul."[4] It shimmers behind everything we see and do. We imagine carrots before
cooking them for dinner, a bath before we bathe. And in literal
terms, our imaginations—saturated with memory—descend from
visual cortex in bundles of neural tracks, meeting incoming signals
in the center of the brain, where sensations mix with imagination
and are matched and filtered. James Hillman says, "In the beginning is the image; first imagination then perception."[5]

Jean Paul Sartre, in *The Psychology of Imagination*, begins a profound inquiry into the nature of the imagination by making two
points. For the purposes of his philosophy, he carefully distinguishes between "the image" and the perception. The image is
what arises in the mind's eye. The perception is the observation of
the material world, those things we literally see, touch, taste, and

smell. The first point Sartre makes is that the perception, the way
we might see and apprehend an object, is infinite.

> In the world of perception every "thing" has an infinite number of
> relationships to other things. And what is more, it is this infinity
> of relationships—as well as the infinite number of relationships
> between the elements of the thing—which constitute the very
> thing. From this, there arises something of the *overflowing* in the
> world of "things": there is always, at each and every moment, in-
> finitely more than we see; to exhaust the wealth of my actual per-
> ception would require infinite time.[6]

Sartre's second point follows, with considerable significance, upon
the first. He suggests that the image, that which is created and
held in the mind, can never be greater than one's actual experience: "No matter how long I look at [a mental] image, I shall never
find anything in it but what I put there." In Sartre's view our image-
making capacity is limited by the degree to which we actually observe—and soak into the realm of consciousness—the myriad
possible perceptions of any given thing. He views most of us as
suffering from what he calls an "essential poverty" of imagination.

Although Sartre's position is apparently contradictory to Hillman's notion that our view of the world follows our imagination,
Sartre recognizes the complex interaction between imagination
and perception. What matters, according to Sartre, is "the position
of consciousness in interpretation."[7] He asks, "Where is consciousness positioned?"

I think of the modern mind and ask myself, is our consciousness
cast out upon the things of the world, projective and perhaps insistent, dominating in its image-making assumptions, stuck on a

preconceived and personal view of the world? Or, on the other hand, is it surrendered, resting inside of us, quiet and receptive? Does our imaginal consciousness cast a flavor of good will, both offering something to the world and allowing the world to be as it truly is? Are we receptive, passive, or projective? The gradations are subtle, but the visionary makes no mistake about the energetic stance with which she views the world. The visionary, in the sense that I am suggesting, knows where her consciousness resides along this internal and slippery continuum. She is not analyzing her "position in consciousness," but simply knows—from a flash of introspection and a cultivated, energetic form of self-knowledge—the degree to which she is receptive, projective, or co-creative. In Sartre's view, the slippery continuum is the imagination itself, implying that the visionary knows, in any given moment, the relative powers—the degrees of influence—of her imagination.

Imagination comprises close to half of any perceptual or psychological cycle. In the circular calculus of the psyche, imagination is a dynamic element, a free-floating radical. It is the shape-shifter, the alchemical cauldron, a treasure chest lying in a sea of possibility, and the house where coyote lives. It is unruly, its influence on the perceptual moment fluctuating, following the lead of strong desires and strong habits. Webster's dictionary tells us that imagination is a "creative talent or ability"and "resourcefulness." But the words "creative talent" imply that our ability to imagine is limited to the gifted few. They suggest a sad reduction of our collective imaginations, a culturally shared loss of imagination—our universal powers to co-create minimized to mere "talent." Similarly, the word "resourcefulness" strikes me as an exceedingly minimized version of shape-shifting. We become resourceful, a form of manipulation, when we have *lost* our powers to shape-shift, to apply potent doses of imaginal consciousness to the perceptual, co-creative moment. Shape-shifting begins with—is most responsive to—what truly is, shaping the world rather than dominating it. We shape the world with our imaginations. From such a perspective, both "creative talent" and "resourcefulness" imply that we have unwittingly diminished the power of our imaginations.

David Orr, a conservation biologist and longtime environmental educator, believes that there are two obstacles to fully experiencing what E. O. Wilson has termed "biophilia," the innate "love of life." One is denial, or what I have referred to as "numb and not-noticing." The second major obstacle is our lack of imagination. Ecopsychologist Robert Greenway says, "Imagination is what's left over after crashing around in cars, being sensually abused by Wal-Mart, or entrained by TV." In other words, by the end of a typical American day, there's not much left. We even seem to forget that we have an imagination, that it is a power, like vision. We act as if we no longer desire wholesome fantasy, a sense of potential and adventure in the invisible and imaginal realms, our imaginations so overfilled by media borne on waves of commercialism. In consumer culture, no doubt, our imaginal capacity is most often filled by what we *don't have*, leading us to forget about the co-creative powers that we *do have*, the powers we share with all we see and touch and feel. We seem to have forgotten the gift of the imagination, believing it to be essentially unreal, or at least "unrealistic"—without reason.

The primary function of the imagination is to bridge between multiple dimensions of ourselves and, more fundamentally, to bridge the gap between dualities, including the perceived gap between inner and outer landscapes. Sartre's central point regarding the nature of the imagination is that it does this bridge work. By bridging between ourselves and the things of the world, imagination, as if gluing the world together for us, creates meaning. We merge the sight of an old oak with our images of what the world must have been like at the moment of its sprouting. In our imaginations, we discover a facet of time, the way it unfolds backward across the history of a land. We imagine the land with a historical perspective and begin to wonder at the plants that grew there and the animals who lived there, the fact of a wooded hillside before houses and cell phone towers. We see an image enriched by time, perhaps embellished by the stories of our families living there as they grew and shifted over generations, with aunts and uncles, and the stories of their mothers and fathers, woven into a tapestry of connectedness. If we had actually grown up in such a world, we would call these images our memories. We would see photographs of cooking with an aunt or fishing with a grandmother. But for many of us, such connectedness with family and land never happened. If such images are reminiscent of the kind of world we wish for our children, we must bridge the past, present, and future with our imaginations.

Mangham's comment about the nature of the imagination, with regard to the visionary, was that we cannot have a vision for the future without an expanded sense of time, without a sense of history. A sense of history, he said, was easily cultivated in small, tribal, or close-knit communities where everybody knew everybody else—their birthdays and initiations, their grandmothers and the stories that went with each one. With such a perspective, it was easy to feel the impact of time on one's own life. This phenomenon of cultural continuity lent itself to the imagination, to envisioning what could be. Our imaginations were stretched. Mangham ended our conversation about time and the imagination by saying,

"Without an expanded sense of time, and especially of the past, we are unable to imagine the future."

A finely tuned imagination has the ability to hold and embellish an image. Color, depth, and precision—the details of any image—can be added by a refined imagination. Such an imagination draws from the physical world, from the past and present, and from the periphery of consciousness. I recall Neal Mangham's grandfather teaching him to look with his peripheral vision while hunting. At the edges of our visual field and at the edges of our consciousness, the world is almost but not quite known. The edge is where our known experience becomes flavored with an unknown wildness, the free radical. In alignment with wildness, imagination becomes both much less predictable and forward-stepping into the realm of possibility.

Like vision, imagination penetrates. As twin powers in the act of searching out, exploring, and revealing, vision and imagination sweep between internal and external landscapes, as if continually scanning. The power of imagination is directed internally. Our sight, the other power, dives into the external landscape. But of course this is simplistic. There is vision in the mind and imagination in the forest—there, between the branches. Vision and imagination co-create one another. Imagination is continually updated and informed by what is seen and heard, by the sights and sounds. I look to clouds, catch a glimpse of a giant frog in the sky, add the image of a sensational toad, and the sky frog becomes more detailed, somehow enhanced. Where does one's imagination begin or end? Where does one's vision begin or end? Sensations flood and reach into networks of neurons, touching the stories held

there, then thrust themselves back into the world, infused now
with a personal twist.

Absolute-threshold experiments in visual science are designed
to identify the threshold between seeing and not-seeing, the very
edge of seeing. Although the results are statistical averages that
delineate a specific threshold, the experience is ambiguous. Dim
lights of varying intensities appear randomly within a dark field.
The subject reports the location within the visual field—the left or
right side of a screen—where the light is being shone. During
some trials, the light is too dim to be seen. Reporting the location
is guesswork and falls to chance; the norm for identifying such
subthreshold light intensities is 50 percent correct. During other
trials, with higher light intensities, the subject may see a faint
shimmer or distinct glow and have better results at reporting the
correct placement of the light. With many subjects and many av-
eraged responses, the line between seeing and not-seeing is thus
delineated. But as a subject in such experiments, I found myself experiencing an ambiguous zone between actuality and imagination.
Was that shimmer real or imagined? I often could not tell whether
I was actually seeing a shimmer of light or just imagining one, and
I could not sense a line between seeing and not-seeing. Rather, I
experienced a zone of *maybe*. Within that liminal zone, imagination, intuition, and several photon-waves of light on the very edge
of seeing were indistinguishable from one another.

For James Hillman, "the imaginal" refers to a seamless realm of
seeing and imagination and to the source of psyche or soul. Like
Carl Jung, Hillman suggests that psyche is image—that psyche is
constituted by the images displayed in the world and received,
carried, and regenerated in one's consciousness, fantasies, and
dreams. In *A Blue Fire*, Hillman suggests a "psychology of soul that

is based in a psychology of image."[8] But in such a psychology, soul or psyche is larger than the individual human soul. In Hillman's view, soul is not limited to humans. It is found in the world, in the images, in the "self-display" of Raven, Rock, or Canyon. In David Abram's view, the "intelligence" of a place is a particular instance of the soul of the world, the *anima mundi*, and is experienced through the sensory body. The soul of a place is the way a particular dense forest, for example, seeps into and informs a receptive human psyche with its self-display, its images, its specific frequencies and vibrations. The coupling between the images offered by the world—the soul of the world—and the human heart, in Hillman's view, constitutes the imaginal realm, the imagination. Imagination thus has the power of blending world and heart and soul, of "coordinating" the powers. Hillman writes,

> Each image coordinates within itself qualities of consciousness and qualities of world, speaking in one and the same image of the interpenetration of consciousness and world . . . The imaginal intelligence resides in the heart: "intelligence of the heart" connotes a simultaneous knowing and loving by means of imaging.[9]

The world soul is found within the images, and the imagination is constructed by these images—by what we see, hear, and feel, by the "words of the land," and especially by a resonance, a sense of affinity, with, for example, Raven or Salmon or Bear. We feel the affinity—a subtle welling up of the heart in response—after stories of each animal are told and retold, after totemic images have resided in our mind's eye, and after direct experience with the chortle and caw of Raven, the immense muscle of Salmon, the raw and seductive power of Bear. With such experience, our imaginations become uniquely human forms of the world's soul, embodying the bridge between ourselves and the very soul of the nonhuman world. In practice, then, our active imaginations are the telling of the world's soul. The feel of such imagination, the phenomenon of an active imagination, is a sense of wonder and belonging.

Plotkin referred to this phenomenon when he described the way imagination creates a deepened relationship to a specific place. He described a favorite mountain retreat for leading Vision Quests, for soul initiations. The name of the mountain, Shandoka, means Storm-maker. Each time he goes there, Plotkin recites Rilke's poem "The Man Watching." The poem describes the "storming of the soul," the internally generated and utter domination of one's ego that both defeats and strengthens us. His memory, infused by the images of Rilke's storm, adds wonder and a subtle sense of knowing the land, of recognizing Shandoka, and of belonging.

According to Angeles Arrien, a cultural anthropologist, the Vision Quest is common to shamanic cultures as a form of visionary practice. For the visionary, questing for a vision is a way of reclaiming the power of one's imagination. A Vision Quest restores the imagination with real life—the squawks and screeches, the sounds we cannot name, the watching and wondering, and the images that naturally arise in the dark of night. Our imaginations become filled with whatever might be sitting in the unknown darkness, just at the edge of our camp. I recall Don chuckling about how going out into the night is visionary practice, and Plotkin's statement that the visionary must "go into the dark." Plotkin meant this in every way possible—to go outdoors on a four-day fast, into the dream world, into trance, or into other altered states of consciousness. The visionary's job is to cross into the unknown so that he may view the world from another perspective, one in which the ego has been set aside, even forgotten. "Something radically unknown must be brought back from the other side," says Plotkin.

While questing for a vision, one is called out from the core, wondering, imagining, and offering one's self. The sensations of soul are remembered, the feeling of being called—even tugged—into the world. It is as if our own soul comes into being as we are pulled by the images, the sights and sounds, of the unknown. On a quest for vision, one's essential purpose and place within a community are clarified and mirrored back by placing one's self within the potent field of wildlife, where life is most dense and most alive, where wildness restores the imagination—where we see the many faces of our own imaginative possibility, our own souls. We draw our alignments on a Vision Quest, noting where we stand, with what and whom we are aligned. This simultaneously clarifies our place and our purpose.

Do we quest for a vision? Do we lend our eyes and ears to each moment, to the raven passing by or to the steep rise of a mountain—our senses suckling the world? Is our imagination filled and nourished by the massive, white form of Mount Blanca, the home of the Dineh indwelling spirits? Do we feel meaning seep through us in moments of looking toward Mountain with curiosity, with a little wonder? Do we imagine the peak of a mountain penetrating the home of the sky gods, as if making the divine realm a little closer to us? Does Mountain represent the edge of the world, the interpenetration of material and immaterial realms, the edge between the visible and invisible, the immanent and the transcendent? Again, this is Mountain as archetype, as symbolic and loaded with meaning. "By attaching *archetypal* to an image, we ennoble or empower the image with the widest, richest and deepest possible significance," says Hillman.[10] We ennoble what we see with our cultivated imaginations.

Like the perception of beauty—which is not found simply in the eye of the beholder—seeing archetypal dimensions in the landscape is a reflection of the psyche, the soul found within *both* the observer and the landscape itself. The reflection of the personal psyche becomes especially apparent with careful observation of the world soul—when one looks toward the familiar mountain, toward the forms with which one is aligned. The changes in one's response, from one day to the next, reflect the changing of the personal psyche more than that of the mountain. In this sense, the landscape becomes a clear mirror in which the individual psyche is seen in what is seen—reflected back with the twined powers of vision and imagination. And in the process, with this way of seeing, the world becomes ensouled.

The power of the imagination is multifaceted. While teaching vision improvement, I occasionally came to class with a kitchen drawer full of utensils and all the related trappings that accumulate in a loose drawer. I spilled can openers and scissors, corkscrews, batteries, Scotch tape, and matches into the center of the table, then asked the students to close their eyes and visualize a pair of scissors. Once an image of scissors had clearly formed in their mind's eye, I asked them to open their eyes and find the can opener. Needless to say, it took them forever, their imaginations already busy with scissors. Next I asked them to close their eyes and visualize a roll of Scotch tape. Upon opening their eyes, I asked them to find the Scotch tape. Needless to say, it took them mere seconds.

We all recognize this phenomenon from our experience. It is an example of the power of an internalized focus of attention—endogenous attention—directed toward a particular mental image. The power of attention, we might say, lights up the neural networks subserving the image of Scotch tape, activating our image-making abilities. Over time the practice of attending in this way, of visualizing—or repeatedly running a signal through the neural network subserving a particular image—strengthens or "facilitates" the connections within the network. And because visualizing a carrot, for example, activates the same neural network as when one is actually looking at a carrot,[11] the practice of visualization makes it easier to see similar things in the sensible, material world. In other words, imagining something makes it easier to see it, just as seeing something makes it easier to imagine. The process between one's active imagination and seeing with clarity is reciprocal and co-creative.

For Malidoma Somé's people, the Dagara of West Africa, thought and reality are closely associated, and imagination and reality are minimally distinguished. In *Of Water and the Spirit,* Malidoma Somé writes:

> To imagine something, to closely focus one's thoughts upon it, has the potential to bring that something into being. Thus people who take a tragic view of life and are always expecting the worst usually manifest that reality. Those who expect that things will work together for the good usually experience just that. In

the realm of the sacred, the concept is taken even further, for what is magic but the ability to focus thought and energy to get results on the human plane? The Dagara view of reality is large. If one can imagine something, then it has at least the potential to exist.[12]

In this sense, imagination is the power "to make happen," to create reality. For the Yoruba people, also from West Africa, the "power to make things happen" is referred to as *ashe*, or "spiritual command," and is the most highly prized of the personal attributes.[13] Ashe is born of rich, imaginal relationships with numerous deities, the *orisha* —the riverine goddess, Yemaya; Shango, the thunder god; and his consort, Oya, the whirlwind goddess. Like the Tibetan Buddhist practices of visualizing complex deities and mandalas, the invocations and dances of the Yoruba orisha hone and refine the imagination, the power to make things happen.

To embellish an image, to make it vivid and to be able to observe the details, to "build anything in one's mind," takes practice. We imagine a scene. We add color and detail. With a little time, the visualized tree glistens in sunlight and branches begin to move. Feelings arise with the image and we look into the image with greater depth, adding greater dimensionality. As we add dimensionality and detail, we strengthen the connections between neurons, making ourselves ready for resonance. The tree becomes more familiar and we easily recognize the shape at a glance, the pattern of ash or cottonwood. The perception of the tree is associated with the visualized form, with a practiced and facilitated network, and thus requires oh-so-little light energy for us to see and feel. We feel a kind of affinity for it. We resonate.

The essential intention of the contemporary visionary is to

strengthen the image that she can "get behind," the one that is backed up by heart and soul. The image she carries will be renewed and refined with each attendant gaze, with each penetrating glance informing her, and with each careful visualization. Over time the image is strengthened, now informing her daily choices, her often unconscious movements toward that with which she resonates. The image she holds shifts her course, steering her as if by an invisible wind.

Imagine the Navajo nation generating clean electrical power. In my mind's eye, I can see the long white blades of wind generators spinning and solar panels soaking up the hot Southwest sun on the reservation. The thick clouds of pollution from the coal-fired electrical plant have disappeared and families are no longer being removed from their lands. The degrading relocation policies of the Peabody Coal Company—their shovels "as big as buses"—are no longer necessary. I can also see the junction of the San Juan River and Highway 160 (also on reservation land), now littered with broken bottles and toilet paper, as a sacred garden—a place for all travelers to take pleasure in a river running through high desert country. There is a medicine wheel laid into sand and there are places to sit under cottonwoods and among coyote willows. Native flowers are lush with river flow and people stop for peace and conversation.

My visions are imaginal, perhaps unrealistic—but not necessarily so. What is most true is that they—and the neural networks with which they are associated—guide my daily choices, perhaps a thousand times each day. In small steps, I move toward a more integrated, more wholesome world.

Reversing the World

It's a shift from seeing a world made up of things to seeing a world that's open and primarily made up of relationships, where whatever is manifest, whatever we see, touch, feel, taste, and hear, whatever seems most real to us, is actually nonsubstantial. A deeper level of reality exists beyond anything we can articulate.

—*Peter Senge*

I stand with my head dangling between my knees, looking at the world upside down. The rearrangement of the world puzzles and pleases me. Colors bump together with a fresh brilliance and I forget names and categories. This un-patterns the world and things reshape themselves before my curious eye. Turned upside down, my imagination wanders freely over the landscape.

The founder of the Soto Zen school in Japan, thirteenth-century master Kigen Dogen, taught a practice that he described as "presenting sideways and using upside down" to break up the resisting "nest of cliche."[14] This was a particular teaching on the nature of knowledge, and a practice for loosening up the perceptual categories, the stereotypes, the expected and programmed view. Presenting sideways and using upside down clearly cultivates a kind of fluidity in the perceptual realm. With a fluidity of mind, constellations of shapes rearrange themselves. The categories—the typical arrangements of things, the bins and stereotypes of our knowing—are set loose. We are no longer limited by our habitual view or framed by our own small inventions, and we may find many solutions to any given problem.

Such perceptual flexibility requires that we participate in the perceptual moment. In practice, we simultaneously deepen our awareness and shift our perspective. We notice where we are attending, shift our depth of field, and move the background into the foreground. We loosen our assumptions and prepare ourselves to see what is not expected, not codified. We call upon our imaginations and rocky shorelines reflected in still water, "presenting sideways," appear as Buddhas and arrows pointing to the north. Clouds shifting with the wind take on both familiar and exotic forms, and our consciousness is stretched. The flexible mind is imaginative, poetic, and open to a new view.

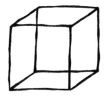

The Necker cube illustrates this flexibility. It is classically used to portray visual illusions, but it simultaneously illustrates the power of the mind to reconstruct the perceived world by shifting one's focus of attention. The Necker cube shows us a simple reversal, created by the placement of our attention, by the way we frame our view. What do we frame as the figure, the left or right facing outlines of the cube? And does the box then appear to be facing left or right? In another classic illusion, we see a young woman or an old woman, depending on what we place in the foreground of our attention.

Similarly, is the redwood forest or lumber in the foreground? Do we imagine clean watersheds and pure air or more stick houses and more money to be made?[15] What is cast into the background, and how does that entice one world into being, leaving the other out of the picture?

The shadows running across the slope of a mountain range, connecting the many folds and drainages in the land, are most often seen as background. If we pull the shadows out from the background, if we give them shape and density—a presence of their own—with the power of our imagination, a new pattern emerges, the figure now exchanged for the ground. The hidden and shadowed places become real, the perception of depth emerges into consciousness, and, if only for a moment, the world is reversed. Similarly, in looking into the human condition, we might shift our attention from the color of skin to the expression in another's eyes. In so doing, our hearts are shifted as well, if only a little. This begins the reversal of the world.

In the complex calculus of the individual psyche, in the way sensations are translated into perception, exchanging figure for ground may not be far from shape-shifting—of turning the world around. Shape-shifting is essentially the power of intentionality brought to bear on a way of seeing. It is aligning one's attention, or "arousal," and one's imagination and thus restructuring both neural networks and perceptual habits, the categorical bins that go along with facilitated neural substrates. Restructuring the patterns of connectivity begins to turn the world around, reversing the trend. Shape-shifting, it follows, is the subtle essence of visionary practice. I see the Navajo nation selling power to the Peabody Coal Company.

There is an art to this. It is the ability to both free one's view from the conditioned and programmed worldview—unpatterning the assumed world—and artfully stitching it back together with the power of a cultivated imagination. Cultivated, in this sense, means informed and shaped by the integrity and wholeness displayed by the visible world. The visionary is receptive, open to the soul of the world speaking through her senses. She attends to the flow of water, for example, to the movement of water in relationship with the rest of the world. She observes the flow forms— the eddies, spirals, vortices, and meanders. Her imagination is fed, informed by the patterns, the language of nature. She sees that water and gravity, together as flow, soften the world—shifting the shape of every river rock. Her observations strengthen a sensibility, the quality of soft persistence, for example, a continual devotion to *what is,* to what is natural and right, the Tao. In relation to water, gravity—*both* the weight and heft of the physical world and an invisible, elemental and universal force—is natural and right. In the Tao, in the flow of water, the visionary sees *both.*

Arthur Zajonc describes an intentional dialogue between the

world around us and our sensory selves as a "yoga of the senses." In Zajonc's view, a yoga of the senses is a conscious and practiced movement between outer and inner worlds that is "profoundly helpful in refining our awareness." He specifically refers to Goethe's method for developing organs of perception tuned to particular objects of visual contemplation and to the Buddhist practice of *kasina*. The kasina practice begins by attending to a plant, for example—"with a deep appreciation, very intently." After several minutes of intense observation, the observer looks away. An emptiness is left in the visual mind, an absence of the plant. Then a "mood or gesture," a kind of afterimage of the plant, begins to surface within one's inner attention. This afterimage is referred to as the *nimita* in the Buddhist tradition. The nimita can be strengthened through practice, by attending to the actual plant again. Then, "when one has saturated oneself with sense impression," says Zajonc, one turns away again, allowing the impression of the plant to inform the psyche, to arise as "mood or gesture." Zajonc emphasizes that "what one sees is the gesture, the inner quality, as it were, of that particular plant. It might have a particular intensity, a kind of exuberance, or perhaps it has a very gentle, modest appearance."[16]

Zajonc suggests that if one can "find the right rhythm—that cadence of engagement, then disengagement with attention to the inner, then re-engagement, then disengagement—there is a growing intimacy with what it is that's before one." Our relationship with the Other deepens; we begin to experience, to feel and know, the qualities of the Other. In conclusion, Zajonc says that "a yoga of the senses or an attentive relationship to the natural world leads in and of itself to a growing synchrony with the rhythms, the textures, the qualities of the world in which we have evolved as a species." This reminds me of my friend Julie, returning from a day at Wet Beaver Creek with a subtle and newfound grace. The day was still and drippy with a soft rain, the creek lush and quiet. She kept saying "It was *so exquisite* . . . The creek was *so exquisite*." It was the quality she emmanated that day.

To gather, hold, and refine an image, and to thoughtfully project that refined image onto the world, is the artistic, creative work of the visionary. The artistry is in fine-tuning our modes of receiving, refining, and projecting images. Envisioning the world as we want it to be runs the risk of strengthening the unconscious habit of projection. The difference between visionary practice and projection gone wrong is one of intention. As a culturally conditioned habit, projection is most often infused with domination, whether conscious or not. We externalize our own assumptions, needs, and habits in the interest of protecting or satiating the conditioned ego. But without our conventional perceptual habits, or an unconscious and egoic value system built into a way of seeing, projection is merely a function of the psyche, another psychological power. If projection is done with consciousness of the fact, and with clarified intention, it becomes a form of responsible co-creation. Hillman refers to a similar, if not identical, form of projecting as personifying. For example, I sit in the arms of a tree when I need to be held. I feel somehow mothered, wrapped in large branches, her arms. Is this projective or receptive? Is my heart engaged or am I visually dominating the world? Hillman describes personifying as imagining things in personal form so that "our hearts can reach them" and as "providing containers for the many configurations of the soul."[17] Most essentially, personifying is using the imagination in the service of the soul, "coordinating the powers" of heart, image, and imagination.

The visionary mode of projection is one in which the internal images, the imagination that shapes and colors the world, are deeply informed by the wholeness displayed for us in the external world. For the visionary, imagination and intention are informed by patterns and relationships, by that which is larger than a singular, egoic interest. Both the soul of the self and the soul of the world are displayed by the patterning held within natural and social systems, in blossoms and mountain streams, in the flow, the Tao—in the very observation of flow. Visionary practice, it follows, begins with observation.

The visionary must be receptive, open to the world, observant, and attentive, informing his consciousness, as if busy in the act of gathering. This includes looking directly at what is, looking to and over the edge, looking into, and cultivating the power to reverse the world, to translate. The practice of the visionary, the shape-shifter, the one who shape-shifts in a troubled time, is a perceptual act.

I watch TV in the Hospitality Inn after driving two days in the rain, contemplating the thousands of dying pines that line the highway in Ohio and upstate New York. Has acid rain caused this devastation, or are the trees dying from some disease? Is this the pine beetle having a heyday? It is painful to see—thousands and thousands of trees in a slow spasm of death. My impulse is to turn away, but I recall Jack Kornfield, a Buddhist practitioner, reminding me that the fundamental spiritual practice is to awaken, to cease delusion. Skillful perception, I remember, begins with noticing what actually is.

I try to understand the dying pines, to perceive the relationships that are operative here. Do the trees reveal air pollution in the rain, now falling hard for two days straight? Is this death spasm a result of the toxic emissions from factories to the north? Are the pines suffering the effects of exhaust spewing from millions of cars, leaving an imbalance in the local ecosystem? There are no answers on the television. Talk show hosts are yammering mindlessly, persevering on our widespread, cultural lack of love. Young unwed mothers, the talk show guests, cry and are blamed for getting pregnant. Everyone seems to be on an excited edge, apparently wondering if they will be called irresponsible whores. No one seems to notice the inherent relationships, especially those between the unknown fathers and their unborn children. The next show is *Baywatch Nights*. I see blonds bounce around the beach, the codified images that commonly color our imaginations.

The next day I continue my drive to the east. I see many road kills, many, many road signs, and endlessly long power lines everywhere. The rest stops are now full-service centers with faxes, many food choices, and overnight mail service—all of it fast like the 75 miles per hour I have been driving all day. I see more land bulldozed, a wide swathe next to the road, soon to be paved for more cars and trucks. I see thirty-five freight trucks pass me in one minute, all of them on their way to the land of more goods to buy and sell. In the midafternoon, I see a deer dashing across four lanes of New York turnpike like an apparition. She did not fit my expectations—her appearance so unusual across the long gray highway—and I nearly missed the sight of her. But there was no mistaking her terror, and my heart leapt in sympathy. A moment later, my sensitivity radically heightened by the sight of her terror, I feel the raindrops hitting my windshield as if striking my own body. I allow—I swallow in sadness—a synaesthetic moment in which I remember that the Others, the Deer, are what make me fully human, a fully sensitive being.

The television shows, environmental degradation, miles of highways and parking lots are the visions that occupy much of the American mind. With little public notice and what seemed to be a thinly veiled jubilance, the largest road bill in history was passed by Congress in 1998. Did we imagine that this law, this public expenditure, meant more truck stops, more trucks, and more shipping of goods around the country? Did we imagine that it would deliver more for us to buy, that it would burn up more fossil fuel and provide more reason to maintain a horrifically high-cost military? It seems we have forgotten both the power and the range of our imaginations. Our imagination offers the opportunity to look at all sides of any reality.

Those who take the vows of the Bodhisattva make a commitment to transform the pain and suffering of the world for all sentient beings. Like the way of the warrior, the fundamental practice of the Bodhisattva is to be fully present in any moment. Kornfield calls such presence "sacred attention." In essence, this means to "pay respects to" all that is, to both the painful and the glorious moments of our lives. For the Bodhisattva, the very basis of wisdom and authentic compassion is to see the truth, to be present enough to see what truly is. Bob Roberts, who directs Project Return, a prison aftercare program in New Orleans, is a contemporary Bodhisattva. His unwavering, unblinking presence with the

pain of others transforms lives. Looking into the face of despair,
Roberts reverses the world. In part, his magic is a perceptual practice.

Tonglen meditation is a specific form of practice for translating
our experience of the world, our perceived worlds. The practitioner, traditionally a Buddhist monk, breathes in the suffering we
witness. Pema Chodron advises adding "texture" to the in-breath:
"Breathe in a feeling of hot, dark, and heavy—a sense of claustro-
phobia."[18] The out-breath is textured by a feeling of "cool, bright,
and light." The practitioner calls upon the imagination to texture
his experience, to add dimensionality in the translation, while
breathing "in with the bad and out with the good." The practice is
a kind of perceptual aikido, a perceptual spin infused with care for
the world. As perceptual translators, we soak in one reality and
spin it back with a strand of beauty woven into the fabric. For example, I look into the face of a now homeless Indian woman, displaced in the process of relocation. I let my gaze linger long
enough so that I am able to carry her image with me. She stays
with me, informing me and nurturing a compassionate sensibility.
Enriched over time, this sensibility surfaces as compassionate activism, as visions of wind generators, as power and economic sovereignty on the reservation, and as a personal commitment to offer
whatever I am able to give. Each of us, informed by what truly is,
may find a way to offer a response, our unique translation between
inner and outer worlds—infused with imaginative possibility and
care. We put our own artful spin on the suffering that was Buddha's
first noble truth. In the act of responsibly translating between realities, our compassion is awakened, our view of reality is expanded, and we experience the great pleasure of giving back to the

world. It is perhaps the most graceful of the "giveaways," our world now infused with beauty.

A series of sitting mudras taught by Erich Schiffmann also emphasizes the practice of translating signals from the external world into compassionate and loving forms of expression.[19] Mudras are typically hand positions that both reflect and further generate inner states of being. The mudras taught by Schiffmann reflect the internal, energetic practice of the visionary. The first mudra is a gesture of opening to the world, "the All." The hands are raised to the sky, the palms open, the thoughts welcoming. After a moment, the arms are lowered, the palms still raised to the sky and held in a gesture of receptivity. In the third mudra, the arms are lowered more, the forearms and palms now parallel to the ground and aligned with the solar plexus, the center of the body. This deepens the reception, the palms still open to sky, to light, to the universal forces. In the next mudra, the hands are raised so that the fingers point to the sky, the palms facing forward and now aligned with the heart. In this position, the heart is visualized as profoundly open, as undefended. The mudra signals a willingness to engage with the world in a selfless, peaceful manner. The fifth mudra begins by lowering the left hand to the knee and resting it in an open position, the fingers lowered toward the ground. The right hand is still raised, fingers toward the sky, and open to the world. The position engenders a sense of both receiving and giving, of participating with the world. The next mudra focuses this experience with intention, in front of the heart. The palms of the hands are brought closer together, still facing forward, both in a receptive position and radiating focused energy outward, from the heart. The final gesture is referred to as the Anjali mudra, or *Namasté*. The palms of the hands meet in the traditional prayer position, in front of the heart. It is a gesture of gratitude.

The tuning fork of the visionary is the gesture. The gestures are relational, moving toward and in response to the nuances of relating with an animated world. We reach out, receive, and give back, continuously engaged in the act of translating between what is and what might be.

The practice of envisioning what might be, the central work of the visionary, is a stretch of the imagination. Here's a stretch: Imagine where you are as green, lush, moist, and fragrant after a deep, drenching rain. In the desert Southwest, imagining this is a big stretch, especially if we have not developed a flexibility of mind. And why should we? We are continually filled with commercialized images that keep the mind preoccupied and cluttered. But this describes the magnitude of the stretch we must make. We must revision degraded conditions, envisioning the kind of world we wish to devote ourselves to, now. We look for green growth, sparkle, the shine of an eye. We visualize a world full of fat forests and wholeness, preserved for all children. This stretch of the imagination bridges between now and then, between what we have learned and what we wish to be true. It's a big stretch. Where do we begin with such an imaginal practice?

We begin where we are. Look about. Where are you? What are the qualities of your perceived reality? Do you see life busting through or degradation? Do you see a twinkle or despair? Are the qualities you perceive seductive to the eye or hard and defended? How many dimensions of this reality do you allow yourself to see? What feelings arise within your body as you look around? Are you pleasured or discomforted by the place you inhabit? Be sensitive and be honest. Look beyond the surface qualities, considering all the relations you see and imagine. Now stretch your imagination, asking yourself what would make this place more pleasurable. How might you offer beauty to the place where you stand, where you live and work and make love? Recognize that whatever your answers may be, whatever paths you take toward beauty, they are all indicated—as if signed by all the relations—by the look and feel of your displeasure. We are most informed by what truly is, and like courage and fear, beauty and the beast are paired powers.

For most of us, where we are is embedded in accelerating forms of degradation. If we recognize that the social and environmental concerns of our time arise from a historical predilection toward not-noticing the myriad relationships within which we are embedded, we wisely shift our focus and attend to the relational world. We look for what has been invisible, sniffing out the relationships and reversing the world. Images of relatedness, of more wholeness, appear. They seep into us, infusing our imaginations, informing us. We learn the way of relationship, the principles that may guide us into a world larger than our small selves, larger than the lonely habits of selfish interest, and larger than the sensibility that goes with them. With practice, we are able to open ourselves more widely and are more capable of translating between what is and what is *possible*—our bodies, your body, a temple for translation.

The visionary must be skillful. Visionary practice includes attending to the relations, the edges and the shifting forms and shadows. We learn to recognize the qualities of our relationships, the flavor of our presence in the world. We train ourselves to step aside, to let the landscape speak for itself in the many languages of nature. We read the signs and symbols, the gestures, the fact of light beaming into a clearing as if God is visiting. We see the wind, once the spirit—*nilchi*, for the Dineh—the animating force that wove the world together, spilling across a mesa, tumbling into desert, spreading over a continent, and sipping on a quiet bay, then entering the skin of the ocean, the sea birds, and all the swimmers. Then shifting, as wind does, it fills the lungs of seals, humans, dogs and cats, snakes and salamanders, the breathers of the world. The world is woven together by wind, invisibly threading a web of relations. As visionaries, we see this.

Neal Mangham concludes a long conversation about the nature of the visionary and the issues of our time by saying, "There is enough." At an environmental college, where an ethic based on both dwindling resources and a recognition of overconsumption is commonly shared, this strikes me as an odd statement. It is such a reversal of my typical view of the world. I ask him what he means. He talks about the apparent conflict between the agricultural needs of the Ndebele tribe in Botswana and the elephant herds that forage in their fields. He paints a picture of elephants and humans in conflict over the same patch of land and says that there is a solution to the conflict, that both can exist. I ask, "Are you saying that the issue is a matter of seeing an alternative to *either* elephants *or* humans? Are you saying that the ecological crisis is a matter of perception?" He looks at me with his head cocked to one side, a little quizzically. Then he smiles and says, "Yes, that's it. Yes, exactly." In my mind, it's a matter of imagination, of cultivating an enlarged and detailed perspective. It is reversing the world, a matter of shape-shifting with the combined powers of attention and imagination.

Shape-shifting is the skilled use of intention with respect to imagination and vision. It is intentionally shifting one's assumptions and projections of the Other. With a slight reversal, we see that the gaudy and oblivious American has a heart just below the surface. We see that the fallen redwood, now dead on a forest floor, supports all sorts of life and is a kind of time-release vitamin for the soil, and that healthy soil is the best of all water filters. Shape-shifting, we see, is unlearning a category, loosening our conditioned assumptions. It requires redirecting our attentional focus and seeing with a relational and poetic sensibility, our hearts resting behind our eyes. It takes practice.

I am in the Grand Canyon again, looking for a vision with depth. As I look around the canyon from Horseshoe Mesa, I am wondering about the vision I already carry, my future image for the world. The Colorado River appears and disappears two thousand or three thousand feet below me. Multiple layers of rock rise thousands of feet above me. I am astounded by where I am—surrounded by deep earth colors and the sky's brilliant blue. I am surrounded by enormous beauty and grateful for the fact of it seeping into me, saturating my soul. I make no mistake about the power of where I am.

My vision is that we will truly see where we are. My vision is that many more of us will awaken, learning to profoundly see and feel the magnificent planet we must now fully recognize as our shared home. Then I imagine more of us being in love with where we are, and that our care for the world infuses our behavior with compassionate forms of activism. I see squadrons of visionaries. They are planting seeds of innumerable variety, reaching out and giving back to the earth. They are "flocking," and reshaping education, carving totem poles and dancing—all with a kind of fevered joy. And, like my friend Liz, they are glowing, radiating with life and transforming the world with spirit.

I envision a world in which many of us do many forms of visionary practice. Ultimately, I realize that visionary practice is ceaseless prayer, and that prayers are profoundly honored visions—arising from careful attention and a refined imagination. Saying our prayers, we see what is possible in our mind's eye and verbalize our most sacred visions. With an honored sense of possibility, we send our visions out into the world, following them in small and caring steps.

Standing here in the Grand Canyon, my vision surfaces with the force of tears. I remember the witches, the millions of (mostly) women burned for direct divination with the natural powers, for having sacred relations with the trees, mountains, springs, deer and hawks, rivers and oceans. I think of the genocide of tribal peoples, of those who traditionally recognized and valued the unknown for its depth. I think of how they made peace with not knowing, with mystery, through the power of ritual and ceremony, through an apprenticeship with animals and places and with refined mental powers. The vision I carry comes closer to the surface now, becoming more clear. I see lovers of the land, those who embody a deep appreciation for the taste and feel of landscapes, the scent and seduction of the places we inhabit. My vision, my eyesight, clears as I write this. Looking up, I see that the canyon walls are sharp and distinct, the colors so profoundly of the earth. They are soft, penetrating, and potent.

I remember *darśan*, the Hindu notion of "getting a view." Darśan is a form of revelation given by Mountain, by the visitations of birds or butterflies, by the sharp edge of a canyon wall, by the horizon. It is a kind of mirrored and reciprocal reflection in which something is simultaneously revealed before our eyes and within ourselves. It is a form of recognition and resonance beyond what can be expressed in English, or perhaps in most spoken languages. Darśan is both an image, worth a thousand words, and the gift of revelation—as if the earth were revealing herself in a moment of mutual respect and intimacy.

Now I remember last night's dream. I dreamt of a message written across a brilliant blue sky: "I do not wish to change the world. I wish to change in response to the world." Then I dreamt of a thick and influential book, like the *Diagnostic and Statistical Manual*, the bible of psychopathology, only it was a manual of psychological health.

Green Tara, the goddess of compassion, sits on my desk. She is a small statue carrying the grace of generations upon generations of Tibetans in worship. She wears a headdress of jewels and a flame rises from her crown chakra. She sits nearly cross-legged, with one foot extending into the world. Her left hand is raised—her fingers pointing toward the sky—and open-palmed in the mudra of reception. The other hand rests on her right knee, also open, in a gesture of offering her intention to the world. Her face is composed, her eyes gently open, her third eye marked with a small circle in the center of her forehead. Green Tara is an image of compassionate activism. She receives the world's energy through her left hand, translating that energy through her presence and embellishing it with her equanimity. Through her right palm, she offers this now transformed energy to the world. She is a subtle shape-shifter in ceaseless prayer. She is an image of visionary practice. In any language, she is an image worth at least a thousand words.

Chapter 7

Looking for a Worldview

I'm not so interested in ethics or morals. I'm interested in how we experience the world.

—*Arne Naess*

Man is not himself only. . . . He is all that he sees; all that flows to him from a thousand sources. . . . He is the land, the lift of its mountain lines, the reach of its valleys.

—*Mary Austin*

Now Where?

> Just as we are "scheduled" to mature sexually—to enter puberty,
> enter menopause—we should be capable of a stage of percep-
> tion that accommodates what is happening to us. If we can not,
> then we have a crisis.
>
> —*Carolyn Myss*

The winding watercourse of the marsh appears and disappears behind waves of fog. It is all very much like our yet to be seen and refined early-twenty-first-century worldview. I catch glimpses of it,
and then it slips away.

The notion of "worldview" has been currently popularized by
a growing recognition that we need to shift ours. Our modern
view, for example, is infused with the notion of endless resources.
But we are *d*eplenishing the planetary system, upon which we
all depend, at an alarming rate. In 1998 alone, 40 million acres
of forests were burned. In Florida, we are currently turning over
200,000 acres of wildlands to development every year. This
overconsumption is ultimately based on our way of viewing the
world.

The tenacity of our modern Western worldview has become exceedingly apparent. It is the worldview most of us in the industrialized world keep living by. It sorts and filters our reality,
determines what we see, and sticks in our minds—or more precisely, in our neurons. Some would say our modern worldview is
"sticky" because we are quite comfortable with the illusion of absolute scientific knowledge, believing that the universe works ac-

cording to known principles and predictable outcomes, and because "stuff" has apparently seduced the soul's hunger for intimacy.
No doubt, it is difficult to disavow the familiar and known for a
world defined by the unpredictable territory inherent in interdependent reality or to give up material comfort for a hint of soul
satisfaction from the invisible realms. The heart of the matter is
fear. Most of us fear change. And fear blinds us.

I'm looking for something still unknown, a new worldview. At
the moment I'm looking down into Sycamore Canyon. From here
I see a vast stretch of open land sparsely populated with piñons
and juniper. Great walls of rock, red and salmon and buff-colored,
rise in the north, the red rocks eroded and spilling into hundreds
of side canyons. The buttes and pinnacles and high ridges are lit
up and golden now in the early evening light, just before sunset,
and the wind whistles through me and the tough little pines and
junipers growing out of this ruddy sandstone slope. Behind the
west-facing trees on the slopes below me the shadows are growing
deeper and darker, the brightness brighter in contrast. It is silent
here, other than the wind—and the sound of a gun going off every
minute or so, sometimes three or four times in a row with the stac-
cato *rap-rap* of an assault rifle. Now six shots rattle through the
canyons, shattering the silence, the air broken open as the sun
sets. This is my current view of the world. The shadows grow
darker in this magnificent canyon. The wind comes up; the sun
goes down. These are constants, unchanged since the Anasazi
hunted and gathered and ate and slept and mated here. But guns
splinter the world now, the once thick and tall grasses are over-
grazed, and I am a deer in hiding, with few places to hide. My fear
grows thick.

Worldview" refers to how we literally view the world. Although this may sound simplistic and circular, we seem to miss the functional significance of such a simple statement. We acknowledge that the insights of Darwin and Einstein served to shift our collective worldview, and we make frequent references to changing worldviews, but rarely do we consider how a shift in worldview influences the focus of our attention and consequently what we actually see. And it is not common knowledge that the focus of our attention has the power to shift and change the way we filter and shape reality. As we have seen, our attention influences the structures of the neural networks that subserve our ready categorization of sensory information. The implication is that our worldview is constituted by our predominant experience, by our habits of perception and the focus of our attention. If we are most exposed to flat surfaces our eyes become unaccustomed to subtle gradations in depth, our gaze less likely to wander into the distance or our souls to know the look and feel of depth. So, too, the neural networks underlying all of this do not develop. With our conditioned worldview functioning as the perceptual filter, it further determines our experience, reinforcing itself and our anticipated perception of the world.

The Toltec notion of worldview, as taught by Carlos Castenada's infamous Don Juan, is that it is an inventory of our experiences, acting as a set of filters and mirrors. These filters determine what enters consciousness. From this perspective, our worldview makes us invulnerable, protected against contesting or uncomfortable views of reality. Further describing Toltec perceptual psychology, Ken Eagle Feather describes a worldview as "a principal force in shaping perception." He adds, "By participating within a worldview, we gradually align our energies with that world. As a result, we actively condition ourselves to what we can and cannot see."[1]

The task of any worldview is to make sense of the world. It does so by sorting and sifting through a vast universe of possibilities. We sort and bind together the bits of information that are salient, that have somehow captured our attention. Within the modern Western worldview, material things tend to capture our attention; catalogs arrive in the daily mail, billboards beckon, and whatever is sensationalized and "new!" assaults our senses while shopping.

But no single worldview is comprehensive enough to embrace the totality of the world. And when we are on the inside of any worldview, it is difficult to see a different side of reality—our very neurons being conditioned daily. We forget the realms of possibilities. To see another reality, we must make a point of looking well beyond our assumptions and habits. I think of it as first peeking around the Western lens, then, having had a taste, simply and boldly stepping to the other side.

I no longer trust the worldview conditioned and reinforced by mass culture. The assumptions that constitute the modern Western worldview are not true to my direct experience, nor do they truly represent the interdependence observed in physics or ecology and honored by every mystical tradition, nor do they offer satisfying answers to current ecological conditions. And yet we keep insisting on more answers from the same worn and weary worldview that has cultivated the mess we're in.

Fortunately, it is this very mess that seems to be shifting many modern minds into postmodern consciousness. It is this very mess that has unsettled the modern mind, that is rendering the collective psyche increasingly ambivalent and uncertain about the assumptions we have inherited. This suggests that the time has come for raising fundamental questions. David Kidner, a thoughtful psychologist, asks why psychology is "mute" about the environmental crisis.[2] This question is rarely asked; yet because human behavior drives ecological degradation, this is certainly a fundamental question. From my perspective as a visual psychologist, Kidner's reference to "mute" may as well be a statement about our collective "blindness."

The ecological crisis reflects a crisis in perception; we are not truly seeing, hearing, tasting, or consequently feeling where we are. Our blindness has tremendous implications for the quality of relationship between ourselves and the "more-than-human world." We do not fully perceive the color and vibrancy of the "ten thousand things," the streams of influence that shift and nudge us into recognition, into seeing the myriad forms of meaning that are displayed for us in every glance. We miss the juicy pleasure of being part of, deeply participatory and inevitably embedded. We suffer from a kind of lonely existentialism, what Sarah Conn, a clinical psychologist from Harvard Medical School, calls "global malaise." "In short our problems stem simply from not seeing,"[3] says Steve Hagen, a Zen priest, in an article entitled "Just Seeing."

As we have seen, our collective blindness reflects a confluence of conditions that envelops us without consent. On the surface, it seems as if we have little choice but to collude with the forces of cultural blindness. But given an analysis of our collective trajectory toward personal and cultural destruction, there is no doubt that the selection pressure is on, and that the evolutionary adaptation we now need is a shift in consciousness. In fundamental terms, our evolutionary challenge is a matter of perceptual development, of perceiving the fact of living within an interdependent reality.

This is no easy task. The canon that our Western worldview posits is that the healthy, well-adjusted adult is autonomous and independent, not interdependent. But in the context of our now long-standing recognition of interdependence—since at least the advent of quantum physics in the 1940s—we are beginning to see that a radically individualized notion of who we are is the antithesis of health and well-being. The alienation and fragmentation that flavor our lives, our feelings of uncertainty, of being somehow displaced and depressed, all suggest an incongruity between our common definition of health and the individualistic mode so common in Western culture. Although the individual is central, so too are communities and the ecosystems within which individuals are embedded. The self, the community, and the ecosystem are nested hierarchies, with one level of organization embracing and including the next. This is an image, a model, of relational reality.

Between individualism, reductionism, and an emphasis on objects for our use, our Western worldview is tightly woven together, a model of reality that affirms itself from several angles, and with several constellations of assumptions. Add in the belief, stemming from the canon of science, that truth can be known in absolute terms, and our collective filter becomes rigid, making our worldview especially self-affirming, almost impenetrable. As a result, our predominant stance is often absolutist and defended against alternative views of reality. At the extreme end of this, we become fundamentalists, ready to define right and wrong for anyone who does not yet know Absolute Truth. I am not saying that the Western worldview is fundamentalist, but rather that it carries such tendencies, and that it does not encourage a receptive stance. As a consequence, we too easily miss the signals, the signs for adaptation, for coevolving with the world-earth system as it is now, today. But reading the signs is a matter of survival, and the signs are immense. The signs are patterns of relatedness. The patterns are sometimes merely outlined like the many contours on a map, but they are always found in the natural world.

Contour lines depict the depth or dimensionality of territories. They help us to see, in depth, where we are. We might say that the contours of our daily lives delineate the territory of fossil fuel use, a territory that remains illusive. The lineup at the gas station, like a contour line, delineates geographic, economic, and shared psychological landscapes, as if the dimensions of a territory we might call international relations are coming into focus. Like extrapolating three dimensions from the contours drawn across a two-dimensional map, when we look at the line at the gas station we are nearly able to perceive the relationships between the use of fossil fuels and the death of Amazonian children. But then, perhaps with a dose of denial and a sense of relief, the perceived relationships slip away.

We know the effects of excessive fossil fuel use. As if teased by the ugly god of excess, we see these effects dancing on the edge of consciousness, on the evening news that enters our living rooms, on the talk shows we listen to while caught in traffic, and in *Time* and *Newsweek*. The results of our craving for fast cars and freedom include cultural decimation, the unnecessary illness of South American children, and whole populations dying because we want gas-guzzling supercars and disposable plastic goods. We begin to realize that in the Amazon basin, the arrival of large oil companies necessitated the pacification or removal of tribal peoples still living on lands rich in fossil fuels, like the Yanomami. The upshot is that they are now struggling on unfamiliar urban streets, or with oil-slicked water, or fighting against previously unknown epidemics like tuberculosis.[4]

We know the story; it is imperialism. If we read the signs, we see that the heavy consumers of the world are perpetrators of planetary degradation. The relational perspective also includes recognition of the ways in which the overconsumers shy away from direct relationship with the earth, and from confronting their (our) collusion in the raping and polluting of the planet. This renders their (our) relationship with the earth almost pornographic—essentially without genuine relationship—casting nature as the pseudo-seductive background for a fire-engine red jeep Cherokee.

Look, we do see the relationships here. There are many, reverberating between the world of Amazonia and our sport utility vehicles. In this case, the relationships are essentially degrading. As a result, we are quickly slipping toward the end of cultural diversity, with over ninety tribes disappearing in the Amazonian rain forest since the turn of the century. Without looking, with the refusal of our attention, we are facing the end of the world as we have known it.

The twenty-first century looms ahead without a clear, shared vision of life on the planet. Our current collective relationship with the natural world holds little promise. It is denial that most fully flavors our relationship with the earth—we look away. But it is not only the planet and particular cultures that suffer, nor is it simply the soul. The body—our bodies—suffers. The senses remain unsatisfied, sex is sensationalized, and intimacy has been a struggle for generations. Furthermore, without a presence of mind—that is, without mindfulness and a refined imagination—we largely project our view onto the world, reinforcing ourselves in unconscious defense. This makes our worldview even more rigid. This rigidity limits our adaptive power—the evolution of our consciousness.

A more fully evolved consciousness, one adapted to current world conditions, must be inclusive and aware of the relationships, the web of life. Such consciousness must include the recognition that the fabric of our lives, the tapestry of life on the planet, is torn at the edges and ripping at the seams. It must recognize that it is now mending time, time to stitch the fabric back together with whatever we've got, with whatever will do the job of reconstituting wholeness and integrity.

Thomas Berry, author of *The Dream of the Earth* and a theologian of remarkable generosity and insight, suggests that human consciousness is unique in its ability to bring existence and wonder into awareness, to notice and to celebrate the beauty of creation, and to revel in the grandeur of life on the planet. Consciousness infused with wonder and celebration naturally engenders generous behavior. In a generous mood, we open outwardly, as if giving and receiving in a singular gesture. In so doing, we more readily see where we truly are and naturally become more deeply integrated, more at home in the world. With a little simple psychology, it follows that we become less guarded or fearful and thus more capable of loving. We become less lonely, more fully engaged in the relational realm, and more strongly woven into the fabric. Thus we stitch and weave the world together. In this way our consciousness truly does the mending, and truly makes a difference.

Ultimately, we are being called upon to develop an ecological consciousness. Surrounded by various forms of fragmentation and exploitation, it is imperative that we develop a systemic perspective, that we recognize global interdependence and develop a deepened understanding of relatedness, an ecological worldview. We must know a relational world like the back of our hands, and then, "like a black cat changing colors," we must do the magic of shape-shifting, reversing the trend toward degradation of planetary life. This is possible from wherever we are, from wherever we stand.

Ecology and Perception Beyond Reason

> There is still another sense, *reason*. This sense is probably incredibly junior to intuition, but few people think so yet. It would help if everyone would look hard at the progress reason has wrought.
>
> —*David Brower*

I stand in a wide sandy wash in Sycamore Canyon. Other than my own, there are no human tracks in the rosy sand. The most significant traces of animal life are bobcat tracks and coyote scat sitting on rocks. I'm a little lost in a maze of canyons and washes but more concerned with finding my new worldview than with finding my way back to camp. I'm on a hunt for health and beauty, wholeness and integrity, and I'm certain to find it here, in the backcountry, out beyond reason. I feel like an outlaw, knowing that I am bending unspoken rules. Who ever said that health and beauty were to be found in rocky canyons, or that answers to the problems of modernity would be found in health and beauty?

My many references to a postmodern age reflect my belief that we are necessarily plummeting into an age beyond reason, beyond the heavy rationalism that has so typified the modern era. We must bring more than our typical habits of rational thought to the issues of our time. In this time, *our time*, the forests are disappearing at a radically unprecedented rate, one third of all fish species are threatened with extinction,[5] and it is conservatively estimated that one thousand species take their last breath each year.[6] For the sake of life on the planet, this *must be* an age beyond reason simply because the age of reason has failed to respond intelligently to the vast implications of contemporary human behavior.

Imagination is thought to be nonrational and is therefore not to be taken seriously in discussions of ethics. With a little reflection, however, it becomes abundantly clear that our imagination informs our view of the world considerably, and thus our behavior toward that world. We seem to forget that ethical behavior, like all our actions, is irretrievably twined together with imagination. For Kant, true moral laws metaphorically reflected natural laws, requiring an act of imagination to perceive. Moral insight, according to Mark Johnson, author of *Moral Imagination*, requires a kind of empathetic imagination, "the capacity to sensitively 'take up the part of others,' and the ability to envision constructive actions."[7] In relation to ethics, imagination, and the ecological conditions of our time, the generative capacity of imagination is perhaps most significant. Lewis Hyde, in *The Gift: Imagination and the Erotic Life of Property*, says, "Once the imagination has been awakened, it is procreative: through it we can give more than we were given, say

more than we had to say. . . . A work of art breeds in the ground of
the imagination."[8] We live in times that demand a work of art; a
skilled, imaginative, and psychologically sophisticated response;
and an ecological worldview.

The word "ecology" refers to the study of interrelationships. It
comes from the two Greek words *oikos* and *logos*. *Okois* refers to
"home"; *logos* to "the study of." The etymology of ecology reveals
that the word is ultimately, or originally, about knowing our home
and the relations that live there. But do we truly see *where we are?*
Are we tuned to the nuances and novelties of the places we inhabit—the birds coming through, signaling a change in weather?
Does the fact that few snowy egrets appeared last spring signify
the toxicity of the Mexican lake where they rest in winter? Do I
note that the lake receives the waste from a leather factory, and
does this signify the success of NAFTA, of the trade bills we somehow agree to, implicitly or otherwise? And do I miss the pleasure
that runs through my body when several young egrets take to
flight? Is the extinction of species also the extinction of experience?

An ecological worldview is a set of assumptions that structure
and color the world in relational terms. By definition, these assumptions are life affirming as opposed to ego affirming, and
process is assumed to be of more interest than individuals and
things. From an ecological perspective, the world is constituted by
multiple processes, continual change, and patterns of relationships. Living creatures are not isolated organisms, but are seen as
tied to others and to their physical locations. The roles and activ-

ities of organisms sustain life processes and dynamic patterns and are consequently seen as more important than relatively static forms of identification.

Interest in the bounded, autonomous self diminishes when we adopt an ecological point of view. Because an ecological perspective is by definition more highly integrated, our attention shifts to our relatedness within a relational domain. Cultivating a relational perspective means that we more easily recognize wholeness. Like the smooth slippage of tide, this recognition of wholeness easily translates into being more certain of our own wholesome choices. And choice, we must remember, is the unique and (perhaps) ultimate power of being human. Given this sense of empowerment, egoic interests fade from view.

But the relational domain is rarely well defined. It is always changing, has messy boundaries, and does not provide absolute answers. On the surface, it threatens the ego with dissolution and can be discomforting, as if our identities are being taken from us. For example, I am often criticized for using the term "we." Students, in particular, at the age of twenty-something, are forming an adult and individuated identity and do not want to be cast as anything that they do not personally identify with. I grant them their sovereignty and rightful development phase. I also acknowledge that when I use the word "we," I run the risk of overlooking a host of ethnic, political, and gender differences, and that diversity is fundamental to the healthy functioning of any ecosystem. From a relational perspective, the differences serve the interest of "we," *and* a singular profile of who "we" are, as a collective form of psyche, is essential to understanding the systemic and anthropogenic influences that shape life on the planet. Given a focus on a larger whole, the ego and personal identification ideally become participatory with "we," not forgotten. Such a perspective is not either/or; we are *both* individuals *and* embedded in larger systems of influence.

Our worldview is shifting from independence to interdependence, from modern and mechanistic to postmodern and relational. Within psychology, this shift in worldview appears as a change in the focus of our attention. The historical focus on the individual and pathology, on the "victim" mentality of the modern era, is shifting toward a larger perception of the wholeness and health of an interdependent self.

The perception of healthy embeddedness, at the core, is a recognition of reciprocity, of receiving and giving back seamlessly. "Seamless" suggests an alternative to the broken experiences, the range of experiences that fragment wholeness. In relation to the self, the broken experience is represented by the contemporary variations on the theme of alienation and of our shared and despairing lack of wholeness. Below our consciousness, broken experiences seep into the core of our identity, leaving us vaguely unsettled, uncertain, unsatisfied, and incomplete. The obvious alternative to this is a restored sense of wholeness, born of the sensations that accompany fully engaged participation with the sensible and sensuous world. This means looking outside of ourselves, not for authority, but for the experience that binds our senses together, that communicates our capacity for a depth of experience, and that weaves us into the more-than-human field of relations. To be more fully integrated is to be more whole; to have integrity is wholesome and healthy. This is an expanded, postmodern definition of health. It includes the fact of being inescapably in relationship with the rest of the world.

But the myriad relationships appear and disappear, as if resting on the very edge of consciousness. How do we see them? How do we hold them in view long enough to get a good look? How do we detect our own deep relatedness to the Others, and how do we refine a relational worldview? I presume that most cultures have long pondered these questions. The answers appear as patterned images and archetypes seen in the landscape. The horizon as an archetype constellates the future and mimes our relationship with it. In a phenomenological sense, the horizon follows us, shifting its visible relationship with us with every step we take. With each new place we stand, our future shifts.

We are further reminded of the relational world by the many depictions of interdependence across numerous cultures. We see carved and woven strands circling the great Mayan temples, embellishing Celtic scriptures, and trimming Chinese furniture. The forms of relationships are seen in the shifting patterns found in Yoruba art, the web of relationships in the image of Indra's Net— an archetype informing the Buddhist worldview. The invisible relationships are also revealed in the alignments between ceremonial structures and heavenly bodies, the sun and stars, or between the cardinal directions, the medicine wheel, and the four sacred mountains. Like a spider's web in sunlight, these relationships shimmer, appear, and disappear again.

Within the traditional Dineh world, the cardinal directions were constant reminders of a relational existence. The medicine wheel, drawn and imagined in relation to the four fundamental directions, joined the human psyche with the landscape, with one's place. The relationship between the land and both personal and collective psyches enriches one's presence and awareness with every 90-degree turn. East was the place of sunrise and recalled the possibility of beginning anew, reminding us that we can always begin again. For the practitioner of the medicine wheel, the powers of the East brought one home, to the present moment. South was the place of the sun at midday, when it is highest and fullest in the sky. It was the place that stirs passion and reminds us of the capacity to love with fullness, brightness, and generosity. The sun disappears over the horizon in the West, recalling the darkness, the inward place of self-reflection and spiritual development. The North was the place of winter, of white snows—reminding us of our elders and of the wisdom to be found in them. It was a place of earned wisdom and disciplined intellect, of the mental arts—the ability to synthesize, organize, discriminate, decide, remember, and imagine.

Perceiving the medicine wheel of Native American cosmology is a relational way of seeing. Learning to know one's place in relation to the symbolic and literal attributes of the cardinal directions is one of the primary lessons in traditional Native American education. Because the landscape continually reflects and informs the psyche, self-knowledge was cultivated by asking, "Where am I?" as opposed to, "Who am I?" Self was thus defined and known in relation to the perceived landscape.

In the Dineh worldview, all beings are expressions of—and actors in—the creation and maintenance of the cosmic order. The Holy People are the indwelling spirits of the land. They differ from the humans, the earth-surface walkers, in that they have complete knowledge of Beauty and are composed of purified energies, the "winds of mind and vitality."[9] Natural phenomena are the homes of the Holy People—the mountains and mesas, the canyons and rivers. In Canyon de Chelly, Spider Rock is the home of Spider Grandmother. A web of rainbows connects the pinnacle of Spider Rock to the canyon walls, weaving the canyon together with shimmering, translucent, rainbow-colored threads. Spider Grandmother taught Changing Woman how to weave, and in this way Spider Grandmother is the very source of the world.

The land is thus "peopled" and a relationship with one's place is a relationship with Others, with archetypes and stories. If self is seen and understood as existing in continual relationship with the land, self is then directly related to the divine. This is confirmed with every thoughtful glance toward the horizon, the cardinal directions, and the mountains and their indwelling spirits, the Holy People. One's self is forever embedded in the landscape, encircled by a mythic horizon.

The perceived world is woven together by the strength of the human imagination. Imagination is a mental art and a primary function of consciousness. Imagination opens the door to the invisible realm, revealing the yet unseen strand between ourselves and the focus of our gaze. Unlike common forms of projection, in which our worldview unconsciously and greatly influences perception, imagination weaves human consciousness into the world and may intentionally co-create the world we see and act upon. And with true reciprocity within a relational system, the imagination further evolves with the world it perceives.

Evolution is never a done deal. Contrary to popular belief, it happens everyday. Our senses continue to adapt, to reorganize through the iterative processes inherent in any organic system. It is the focus of our attention that draws in the influences that shift and shape our worldview, that "pattern up" in our brains. In quite literal terms, as our view changes, our worldview shifts. The way we see, hear, and feel is not limited to habit or heredity. It depends most on where we are looking.

In any one place there are many ways of viewing the world,

representing entire realms of perceptual possibility. In the best of possible worlds, I clearly see where I am. I find my place in the world by looking around me. With what and with whom am I aligned? In the best of possible worlds, we are guided by what we see and hear and feel, by the sensory and sensual interaction with the world that ultimately sustains our every move. In the best of worlds, we are initiated and welcomed into the world by looking, by questing—by looking with care. And sure enough, I am questing for a vision. I am looking for a way of seeing the world that will inform and guide my passion and prayers. All of this, however, assumes good-heartedness, the eye of heart strengthened and transmitting to the soul. The transmission is an image of wholeness displayed by a view of integrity—the world still held together, unfragmented.

Looking for One's Self

The Western conception of the person as a bounded, unique, more or less integrated motivational and cognitive universe . . . is, however incorrigible it may seem to us, a rather peculiar idea within the context of the world's cultures.

—*Clifford Geertz*

We are a landscape of all we have seen.

—*Isamu Noguchi*

At each moment we must simply be what we are, opening onto a larger life.

—*Thomas Berry*

I am both inside and not inside. I am inside a buzzing, blooming world and yet somehow independent, set apart, on the outside of the world by virtue of my own interiority. From the perspective of my experience—my truest source of self-understanding—both are real. Self is a both/and proposition—we exist both inside and outside. But immersed in a social world already defined for me by either/or habits of thought, my sense of self is habitually viewed through a dualistic filter. Under such circumstances, I don't really know who I am; my sense of identity is somehow confused, somehow limited to an either/or interpretation. But if I persist in my search for self, sincerely asking and opening myself to the complexity of the world as if it too matters, or with the idealism of one who is profoundly in love with the world, I come to know myself as idealistic, a lover, passionate. The world, or more accurately, my experience of the world, thus creates me.

Attempts to define, characterize, or even outline the contours of the contemporary self are a tangle of uncertainty. Although Walter Truett Anderson, in *The Future of the Self: Exploring the Post-Identity Society*, suggests three variations on the theme of identity, his basic conclusion is that the contemporary self is fragmented and in crisis.[10] The forms of fragmentation are familiar to us. We radically consume the planet's "resources" as if hunting and gathering for personal identity. We spend several billion dollars a year on various designer drugs to ward off both the truth of our own psyches and the true state of the world, and we are currently building more jails than ever before. The phenomena pointing toward contemporary forms of fragmentation are no surprise. "Everybody knows it," says Anderson.

What we don't commonly recognize is that our descriptions of the contemporary collective psyche, the cultural conditions within which we are embedded, and our own confused sense of self are essentially one and the same. We rarely fathom the iterative power of historical conditioning, and, furthermore, we seem to believe that our selves are somehow neatly protected from the social and ecological disasters of our time. Our notion of self, perhaps still most Freudian in flavor, is commonly skin-encapsulated; self generally refers to "ego," and ego, we find, is comfortably enshrined within the human head. A healthy American ego, we are commonly told, is independent and busy mediating between the wild and to-be-controlled id and the somewhat irritating superego. All told, our common Western conception of the self is essentially "bounded" and sabotages our awareness of psychological interdependence.

Our constructed notions of self, like our perception of the world around us, are heavily conditioned by individualism and materialism. With even a little perspective, history shows itself, as if recapitulated and progressively transformed, in contemporary forms of alienation. The forms of alienation are essentially objectifying and nonsensual in nature and, ultimately, dualistic and divisive. We value stuff and we divorce people. We must remember that it is from this state of conditioned consciousness that we attempt to define ourselves. But this state of consciousness essentially precludes the conceptual possibility that the self is largely constituted by the seamless streams of energy arising in the world and running through the senses.

As an alternative to the radically individualized self, ecopsychology and deep ecology have proposed the "ecological self" as a model of identity. For Arne Naess, the Norwegian philosopher who coined the term, the ecological self models the natural process of maturation in which one becomes increasingly identified with the larger community of all living beings, a process he calls "self realization."[11] At the core of Naess's conception is the notion that we tend to underestimate ourselves by confusing self with ego. In so doing we minimize our potential. Alternatively, when we develop an expanded sense of self—one in which we identify with life itself, and especially with the variety of life forms with whom we share an intimacy—we "see ourselves in others." We see the potential for loyalty in Wolf, for gentleness in Deer, for depth in Ocean.

But there is another fundamental way in which we see ourselves in others. David Abram says, "We may think of the sensing body as a kind of open circuit that completes itself only in things, and in the world . . . I am a being destined for relationship: it is primarily through my engagement in what is not me that I effect the integration of my senses and thereby experience my own unity and coherence."[12] In other words, we know our wholeness, including our sensory and sensuous selves, only through perceiving the Other.

Given a deepened appreciation of interdependence and an inclusive sense of self, we naturally come to care for the world beyond our independent egos as if it were our "self." Our acts of care for the Other with whom we identify then become beautiful acts as opposed to moral acts—the "shoulds" that do little more than linger in the psyche. Identifying with the Others not only serves to widen and deepen ourselves but also to create an ethic born of self-love and enlightened self-interest.

Also with an eye toward interdependence, Paul Shepard refers not to "the self" but rather to "the relatedness of self." He says, "The epidermis of the skin is ecologically like a pond surface or a forest soil, not a shell so much as a delicate interpenetration. It reveals the self extended and ennobled . . . because the beauty and complexity of nature are continuous with ourselves."[13] Frances Vaughn, a transpersonal psychologist, defines the self as an open, living system in an intricate web of mutually conditioned relationships. This is consistent with the self suggested by systems theorist Gregory Bateson. According to Bateson, self is the entire system of receiving, interpreting, and responding to incoming signals—and then receiving again, and again responding. James Hillman summarizes this sense of self by asking, "Where does the 'me' begin? Where does the 'me' stop? Where does the 'other' begin?"[14] From yet another perspective, James Elkins, an art historian, claims that the self is continually transformed by the act of seeing. An image, suggests Elkins, is "a corrosive, something that has the potential to tunnel into me, to melt part of what I am."[15] He further adds:

> There is ultimately no such thing as an observer or an object, only a foggy ground between the two. It's as if I have abandoned the place in the sentence that was occupied by the words "the observer" and I've taken up residence in the verb "looks," literally between the words "object" and "observer" . . . what I have been calling the observer evaporates, and what really takes place is a "betweenness" (for lack of a better word): part of me is the object and part of the object is me.[16]

Given an interdependent sweep between "in here" and "out there," I am suggesting that we are embedded selves, living intimately within a relational reality. By seeing, hearing, touching, and tasting—by ingesting—we become the world *within which we are*. Like breathing in and out, we cannot help ourselves. Whatever we may think about the self, we are most fundamentally at the mercy or blessing of wherever we may be, and of the quality of our experience there. We are truly "where we're at."

The psychological interdependence portrayed above is a simple assessment based on observation and a basic understanding of systemic influence, or mutual causality. As a visual scientist, the reality and significance of mutual causality are particularly astounding with reference to changes in the connectivity between neurons as a function of sensory input; the world—in the form of sensory signals—alters cortical structure. This, in turn, alters categorical perception and subsequent behavior, our action upon the world. Thus is the co-creation of self and world, mutually arising by looking.

In other words, *by looking,* by repeatedly activating neural networks with our attendant gaze, we develop our way of perceiving, responding, and being in the world, and thus we develop ourselves. This portrays the self as an energetically open system, one that sorts and sifts incoming signals in relation to the previous state of the system. The self is thus busy in the translation and transformation of energies, is continually creating and re-creating itself, and is continually engaged in the process of coevolution with the world. What matters, then, is not an identity, as if relatively "permanent," like personality, or "arrived at" with sufficient maturity, but rather the activity, the act, of *becoming.*

The sorting and sifting—filtering in this case—refers to the power of attention. If attention selectively filters the influx of the sensible world and thus determines what we see, then on our end of a co-creative spectrum, it is ultimately our attention that influences the strength of synaptic connections, the emergent perceptual habits, the consequent worldview, and our actions—our agency in the world, our sense of self. *Attention* thus constitutes our being in the world, depending of course on where we are looking. *Where are we looking?* What is it that, daily and over time, constitutes the postmodern self? This is not a mere theoretical consideration. The practical considerations of this question are immense; who we are is crucial in the context of the accelerating use and abuse of planetary resources. Are we identified with having the right stuff, the new dress, the newest sport utility vehicle, the newest anything? Within the human-created, industrialized world of tremendous choice, the shelves are lined with variety ad nauseam. Our choices—reflecting both where we are looking and our assumptions about what it means to be whole and satisfied—have planetary implications.

If there is any single truism about the self, it is that self is a matter of mind. On multiple levels of being, we "decide" who we are. We witness self as "a matter of mind" across a spectrum ranging from "making up our minds" (and thus defining self by virtue of agency) to the deep, historical conditioning of the modern mind, from the cognition issuing from direct experience to the neuronal ensembles constituting categorical experience, consciousness, and subsequent behavior. And perhaps ironically, self as "a matter of mind" includes the Buddhist recognition of no-self, of egolessness. With even an introductory practice in the Buddhist meditation forms of vipassana and samatha, the impermanence of the self becomes apparent; self is revealed as the relentlessly impermanent thoughts and feelings that continuously fill our minds. Self, from the perspective of meditation practice, turns out to be more than experience spilling through mind. If we shift to a Western perspective, it is also worth noting that only the focus of our attention coalesces signals into experience or gives depth and density to the flow of information passing through the mind.

Given a view of the self that integrates systems theory and neuroscience, the question of self, of who we are, is now considered in terms of who we once were (the previous state of the system) and who we are becoming as we gather and integrate new signals in each ongoing moment and over time. "We are a landscape of all we have seen." In an immediate and active sense, we are integrators, mixing and matching the complexity of each past moment with new incoming signals. In so doing, by focusing our attention on this moment or the next, on this flower or that item, we are the creators of self and world. In other words, we choose—and thus we become.

As we have seen, James Hillman describes attention as "attending to, tending, a certain tender care of . . ."[17] Paul Rezendes, in *Tracking and the Art of Seeing*, says, "Caring *is* attention."[18] Coming full circle, we find that this definition of attention is consistent with Heidegger's notion of self. Heidegger coined the term *dasein*, essentially meaning "being there," not simply "to be." From Heidegger's perspective, self and this (inclusive) place now are together as one being; self is its world. For Heidegger, dasein further implied "taking care of." Dasein is an active caring, its essence is care. Like Arne Naess's notion of "beautiful acts," care follows naturally from an identification with that which extends beyond the narrow ego. For both Naess and Heidegger, care becomes the common element of being in the world. From this perspective, *we are* only to the extent that we care; we are what we care for. Andy Fisher, an ecopsychologist, puts it this way: "If we are 'selfish,' that is, if our care is narrowly focused on a 'me,' then our world and our self-sense will likewise be very restricted (and vulnerable). If, on the other hand, our care is wide and 'selfless', then our self-sense will also be wide and deep . . . authentic." Self, care, attention, and the world are an integrated, singular self-organizing system, inseparable, and potentially "wide and deep."[19]

If self is a matter of mind, and "a matter of mind" ultimately refers to attention, then perhaps our search for self ought to be guided by the question: "What do we attend to? What do we care about? What is the experience we choose? Who do we choose to become in a resonant moment of experience? With the gift of choice, *how now shall we become?*"

Richard Nelson, an anthropologist deeply embedded in the Alaskan landscape, illustrates both Heidegger's sense of self-and-world and his own attendant choice in becoming:

> There is nothing of me that is not of earth, no split instant of separateness, no particle that disunites me from the surroundings. I am no less than the earth itself. The rivers run through my veins, the winds blow in and out with my breath, the soil makes my flesh, the sun's heat smolders inside me . . . the life of the earth is my own life. My eyes are the earth gazing at itself.[20]

When we allow the images and sounds, the touch, feel, and substance of the world to enter us, we become the world, with our small, earthly selves included in the mix. This is literal. We eat and become deer. We drink and become the water of the earth. In the case of vision, light enters our bodies in streams of photons. Light mixes with neurons, ripples through cortex, shapes and shifts neural tissue, stimulates the pineal, and releases hormones. Light changes us; what we see changes us. "Light," says James Elkins, "burns into me; it remakes me in its own image."[21]

The embedded self is rooted in experience, dug in and deep. We lose our small "independent" selves and grow in ever-widening gyres of connectivity, becoming and being in a seamless stream of being-in-place. It is no wonder that the most common expression of self among tribal peoples—those embedded in unbroken experience—is the notion of *becoming*. And it is no surprise that identities are often mixed with animals, that the earth is endowed with kinship, that the landscape speaks, that totems serve as intimate guides, and that the land and all the relations are thus valued.

Neither ourselves nor our worldview are static. Like anything resembling or arising from an organic system, both are held within a dynamic and interdependent system. Both change with every influx of energy, every photon of light signaling a transmission. We are embedded and can't help ourselves, unless, of course, we help the Others—those who in turn influence us.

In a final moment of contemplating the nature of self, I look to water, the tide and marsh before me. It moves, like me, spilling, shifting, shifting sand, the edge of land continually eroding. Pebble and stone, soil and sand, slip and tumble into new forms, new configurations. Grass and roots release, die, drift to sea, buried at sea, becoming one, the same. Sand, soil, marsh, grass, marsh, sea, the blood in me; I sweat salt. And pulsing like tide, I become the sea.

For a moment, I become vast, expansive, deep. It is visceral. In the next moment I become gratitude itself, in quiet response to the possibility of such depth.

Way Beyond Worldview

> Look thee from a world of everlasting bliss to share.
> —*Francis Doughty*

My students and I go to the truth mandala. It is a circle laid on the ground in a local ponderosa pine forest. The cardinal directions are marked by stones, sticks, bright orange manzanita berries, and dried yarrow, last fall's bloom. The stones also mark a place for our despair, our sorrow. The sticks hold a place for anger, berries mark our joy, our abundance, and the dried flowers hold our hope. I have asked my fourteen students to speak to the emotions they feel, bringing something representing their strongest emotions regarding the state of the earth to place within the mandala. We hear stories of loss and adventure, poetry and tears, the human heart of the world. Miana speaks last. She is twirling a piece of golden spiral grass in her hand. She says, "Movement is sound." We all wonder what she means. She says again, "Movement is sound," and begins to cry. Although we still don't understand, we all love her for it. She is so very sincere.

Three weeks later, Miana comes to class with a projector, an electric fan, a large white piece of cloth, and her friend Mary Ann. Miana hangs the cloth from the ceiling, about a foot from the wall, turns on the fan, and turns off the lights. Mary Ann begins to play a violin softly and Miana turns on the projector. We see an image of golden spiral grass moving, as if in wind. Then we see another,

Miana slowly showing us one image of grass after another, close up and magnified. Mary Ann's lilting music shifts with the projected images of the curving, elegant grass. The experience immediately seeps into me. I find myself inside the spirals, following the music with my eyes, ears, and heart. It is an intimate look at the world. There are no words to name it, and so I cry a little in the darkness.

Miana turns the lights on, beaming and gesturing as if we all understand. She says to the class, "Now you know why I cried at the truth mandala . . . remember? This is what I saw and heard and felt with the grass. It's visual sound," she says, as if we must surely get it by now. I mention the power of synaesthesia. She has never heard the word but beams brighter, knowing exactly what I am talking about. She tells us that she "sees through feeling." I think, yes, this is certainly seeing with the eye of heart. "It's a *collaboration* with feeling," she emphasizes, "and it's about looking into the spaces. . . . That's where the sound is, in the spaces." She says that the spaces are "the source of the movement." In trying to find words for how she sees, Miana says, "A vibration comes through me, connecting with my body, my insides, my organs, the source within *my* body, making *me move* inside. It moves me. My tears are the movement in me. It makes me want to see and hear again and again and again." Then she adds, radiating, "This is coming home. I become the spiral grass."

I am stunned by her brilliance and think, yes, she's looking with love eyes.

Empedocles, I presume, also saw the world with love eyes. His theory of vision began with Aphrodite, the goddess of love, "fashioning the eyes." He believed that the things of the world were a changing mix of earth, air, fire, and water, the sacred elements rearranging themselves before our eyes. In the context of the new headiness of rationalism and writing, of being able to reflect on ideas written into text, Empedocles' view did not stand a chance against Plato's idealization of the rational mind. But just imagine: What if Empedocles' view of the world was truly born of love? What if he was the theorist recorded in our philosophy books, the philosopher that we all knew and remembered? What if, by some historical chance, Empedocles had been "the greatest influence on the Western mind"?

This book is meant to be a thoughtful attempt at providing a path through the extreme forms of rationalism and dualism that fell out of a long series of thinkers and historical events. It is meant to guide us, as if down a sunlit and shadowed path, through the psychic numbing and chronic, unsatisfied desire to *do something*, to more fully care for our homes, our planet. My premise is that we do many things: recycle and clean our streams, attend to our language and to the way we educate our children, educate ourselves, work to shift our consciousness and to become mindful, and allow ourselves to feel deeply, to care passionately. My own experience has shown me that care for the planet follows from true desire and passion, which naturally follow from seeing, touching, and tasting the beauty of our world. Like any love affair, loving the planet comes through the senses. For most of us, this implies a perceptual shift.

I asked my friend Glendon, an environmental ethicist and writer, an Alaskan backcountry man, what an ecological worldview is, what it looks like to him. He said, "It's love. Nothing less." I didn't quite get it. So I asked him the same question the next day. He said, "It's love. An ecological worldview and love are the same thing." I was still a bit mystified, but James Hillman makes it all clear to me. He says, "We want the world because it is beautiful, its sounds and smells and the textures, the sensate presence of the world as body. In short, below the ecological crisis lies the deeper crisis of love, that our love has left the world; that the world is loveless results directly from the repression of beauty, its beauty and our sensitivity to beauty."[22] This I understand. I understand that nothing less than love will move us enough to meet the challenges of our time.

Donella Meadows is a systems theorist. In 1970 she and a group of colleagues at M.I.T. ran a computer simulation of the state of the world. They published their findings as a book called *The Limits of Growth*. It was published in twenty-some languages and argued about for years. Their conclusion, the source of all the fuss, was that there are limits to the industrial growth model of reality, that we, as planetary consumers, would eventually run out of resources at the 1970 rates of economic and population growth. But since 1970—despite the fact that the population growth rate has declined[23]—our global use of paper alone has increased three times and is expected to double in the next fifteen years. Although this reflects a robust economy, it also forecasts decimated forests, floods, property damage, the destruction of watersheds, desertification, the displacement and degradation of remaining forest cultures, the continued rapid rate of species extinction, and the extinction of experience.

Twenty years later, Meadows and her crew of systems theorists ran a much updated version of the earlier computer simulation. Input to the simulation included figures denoting the number of cars and trucks in the world, the global mileage driven in automobiles each year, and how many drivers have accident insurance. Every conceivable detail was included. In 1990, their conclusion was that we had thirty to fifty years before economic, social, or environmental breakdown. Their interpretation of the outcome of the simulation was, in their view, conservative. But so devastating were their results that they had to "talk about love," says Meadows. As she tells this story to an audience of mostly Harvard psychologists, she reminds us that the people who ran the study were six M.I.T. scientists. "We had never talked about *love* before."

I am sitting in the backyard, bundled up in warm clothes and writing under a full moon. I think of John Lennon singing "All you need is love." To the hard hearts of his time, it was a threatening message and I suspect that his radical campaign for love challenged the defenses we all carry. Love is threatening. It changes us—like light, and scent, and birdsong. It changes us, like seeing the slippage of tide and feeling the subtle sway of ocean, the earth turning and my own heart rising to meet the sweet flood of salt water. Seeing changes us.

If you wish to truly see, says Annie Dillard, "center down." She says, "I go calm. I center down wherever I am. I retreat—not into myself, but *outside* myself, so that I am a tissue of senses. Whatever I see is plenty, abundance."[24]

My prayer is that we "center down," for the sake of all the relations, for all of us. To be perfectly honest—and there can be nothing less—my prayer is that we *get down*, that we get down and dirty. I pray that we lose ourselves while lovemaking with dirt, with the rocks and streams, the salmon who swim there, the coyotes and 'coons, the water bugs and snakes—with the fertile ground of wherever we may be.

Notes

Introduction

1. There are a number of very specific practices, largely based on the work of opthamologist Dr. William Bates. His book *Better Eyesight without Glasses* is the original source of the practices that most served me. The central premise of the Bates Method is that the relaxation of the extraocular muscles allows the eye to return to its natural spherical shape. This, in turn, allows an image to be clearly focused on the plane of the retina. (A nearsighted, or myopic, eye is elongated and an image focused by the lens falls in front of the retina, causing a "blur circle.")

2. In the context of vision improvement, it is worth noting that the ciliary muscles of the eye, the muscles that do the work of accommodation so that we may focus on the near point, are at rest when looking into the distance. More than anything else, it is the relaxation of the eye muscles that supports clear vision.

3. This is a fundamental premise of ecopsychology; the planetary and personal realms are seen as highly interdependent.

4. Robert Michael Pyle coined this phrase to refer to the loss of experience in our neighborhoods—in the vacant lot, the swamp, the places where children discover tadpoles and frogs. There is a specific chapter entitled "The Extinction of Experience" in the book *The Thunder Tree: Lessons from an Urban Wildland*, 1993 (pp. 140-52).

5. D. Abram, 1988.

6. Y.-F. Tuan, 1974, p. 12.

7. R. Abraham, 1992, p. 74.

8. J. Hillman, 1975, p. 122.

1 : Varieties of Visual Experience

1. J. Hillman, 1975, p. x.

2. A. Zajonc, 1993, p. 20.

3. Ibid., p. 37.

4. Quoted in D. C. Lindberg, 1976, p. 10.

5. Saint Bonaventure, 1956, p. 16.

6. H. Smith, 1991, p. 261.

7. Ibid., p. 261.

8. R. Lawlor, 1991, p. 376.

9. Ibid., pp. 315, 381.

10. Ibid., p. 17.

11. D. Abram, 1996, p. 164.

12. J. Highwater, 1981, p. 68.

13. Ibid., p. 61.

14. In *To Summon all the Senses,* David Howes claims that this is why "there is nothing healing about most contemporary Western art." The idea is simply that we do not let art serve us by gathering together the senses so that we become more integrated, more whole, and healthy.

15. T. N. Hanh, 1988a, p. 3.

16. T. N. Hanh, 1988b, pp. 10–11.

17. D. T. Suzuki, 1991.

18. D. Eck, 1985, p. 3.

19. K. Wilber, 1997, p. 290.

20. J. Hillman, 1975, p. 121.

21. This recalls Arthur Zajonc's suggestion in *Catching the Light: The Entwined History of Light and Mind* that "Galenic resonances" refer to "like meeting like." For Ralph Abraham, a complexity theorist, beauty and resonance are equivalent.

22. M. Berman, 1988, p. 18.

23. Ibid, p. 62.

24. T. Roszak, 1972, p. 334.

25. A. Zajonc, 1993, p. 203.

36. T. Roszak, 1972, p. 334.

27. A. Zajonc, 1993, p. 212.

28. Although I am presenting a very abbreviated overview here, the history of science has much to do with our notions of vision. In Chapter 2, I provide greater detail regarding the ways in which rationalism has shaped our ways of thinking about vision and our ways of actually seeing.

29. H. Foster, 1988, p. 48.

30. D. Abram, 1996, p. 49.

31. Ibid., p. 49.

32. Ibid., p. 55.

33. Theoretically, when the "noise"—or spontaneous firings in the visual system—is temporarily quieted by meditating, the signal is rendered relatively more salient and is therefore more easily seen. This interpretation of the visual effects of meditation is derived from signal detection theory.

34. J. Mander, 1991, p. 32.

35. P. Gold, 1994.

36. P. Devereux, 1992, p. 38.

37. Ibid., p. 36.

38. M. Thomashow, 1998.

39. "Chakra" is the Sanskrit word for "wheel" or "circle." Ancient yogic traditions of India describe seven chakras or centers of energy within the body, each active in the reception, assimilation, and transmission of life's energies. The root chakra is located at the base of the spine and corresponds to the color red.

40. The heart chakra is the fourth chakra. It is located in the center of the chest and corresponds to the color green.

41. R. Lawlor, 1991, p. 174.

2: Numb and Not-Noticing: How the Modern Eye Sees

1. J. Mander, 1992, p. 32.

2. P. Breggin, 1994.

3. M. Berman, 1989, p. 25.

4. B. Tuchman, 1978, p. 94.

5. The Black Plague was a tremendously significant event in altering the Western psyche because of its magnitude and rapidity. Everyone witnessed it. It was not limited by class or confined by geography. Paris lost half its population; Florence, three-fifths to four-fifths; Venice, two-thirds. Cities and towns collapsed, fields were not harvested, and sheep, goats, chickens, and oxen also succumbed to the plague in unthinkable numbers. Wild nature took over. With so few people to attend to the crumbling infrastructure, it was believed that the end of the world was being dealt by the hand of a very angry God, and for no apparent reason.

6. R. Tarnas, 1991, p. 225–226.

7. Ibid., pp. 227–228.

8. M. Berman, 1988, p. 17.

9. H. Smith, 1991, pp. 215–216.

10. Ibid., p. 214.

11. Quoted in M. Berman, 1989, p. 23.

12. I am advocating the development of skillful boundaries, a kind of energetic opening and closing of ourselves with intention, a refinement of our notion of black-and-white boundaries. The distinction reveals the way even our notions about the body are infiltrated with dualistic consciousness.

13. T. Nagel, 1986.

14. D. Abram, 1996, pp. 63–64.

15. J. Hillman and M. Ventura, 1992, p. 125.

16. I recall my friend Tom: One day, after his return from five or six months in the Alaskan backcountry, he was in a large hardware store and had a sudden "plastic attack." He paled, nearly doubled over, gasped, and said, "I have to get out of here." He reeled and nearly stumbled through the automatic doors. Although his body was "awakened" by virtue of having been in the backcountry and hence undefended, he paid a price for it; he couldn't go shopping for some time. I tell this story with hesitation, because it is clear that a sensitive body suffers in the mall.

17. E. Langer, 1997, p. 4.

18. M. Berman, 1989, pp. 23–24.

19. C. Glendinning, 1994. p. 64.

20. Ibid., p. 13.

3: Mindful Eyes: Seeing as If the World Matters

1. P. Novak, 1990. p. 8.

2. J. Hillman, 1989, p. 18.

3. In an article entitled "Just Seeing," Steve Hagen describes pure perception as that which naturally occurs before, or beyond, conception. Consequently, "reality is inconceivable," and true seeing is "keeping our mouth shut." In a culture determined to know, keeping our mouths shut—that is, relinquishing certainty, reserving judgment, or minimizing projection—is both an active and a receptive process. To see in this way, we must step beyond egoic habits, releasing assumptions and categories, and extend an open invitation to that which we lay our eyes upon.

4. P. Hejmadi, 1990, p. 71.

5. F. Curtois, 1990, p. 19.

6. Studies spanning a decade measured reaction time and percentage correct and showed large benefits in high-probability conditions and relatively small benefits and no costs associated with low-probability conditions. The fundamental assumption was that probability determined the degree to which a subject was willing or able to attend. All told, an informal survey of the literature and armchair calculation indicated a more or less 20 percent increase in processing efficiency for every 250-millisecond eye fixation. Across several minutes of seeing, the benefits dramatically add up.

7. A. Mack and I. Rock, 1998, p. 25.

8. "Hard-wired" describes those neural substrates that are not alterable by experience. For example, all human eyes (excluding those with color blindness) contain three kinds of color photoreceptors. This determines the detection of different wavelengths and gives us the perception of color. The ability to discriminate color, however, is alterable through experience, or by training. This implies that higher-order processing of color is soft-wired, or "plastic."

9. Varela et. al., 1997, p. 96.

10. Ibid., p. 96.

11. Ibid., p. 87. In 1949, Donald Hebb suggested that learning could be based in changes in the brain that stem from the correlated activity between neurons: "If two neurons tend to be active together, their connection is strengthened; otherwise, it is diminished."

12. E. Clothiaux et al., 1991.

13. This refers to neural tracks that require noradrenaline or acetylcholine as transmitters between neurons.

14. Varela et al., 1997, p. 111.

15. E. Hall, 1966.

16. B. H. Gunaratana, 1998, p. 28.

17. K. Wilber, 1997, p. 283.

18. Varela et al., 1997, p. 22.

19. Ibid., pp. 97–98.

20. Ibid., p. 98.

21. Without the cynical edge, my private theory about ADD is that, like psychic numbing, it has a great deal to do with all that we do not wish to witness; think of the many schoolchildren that know—in their bodies and yet unconsciously—that sitting in rows with thirty-five other children listening to the (all-too-likely) tired schoolteacher recite the lessons of yesteryear is irrelevant to the issues of our time and to the play-world of the child. In other words, ADD represents a constellation of responses to a world that doesn't fit the true human condition. By true human condition, I mean the hunger for relevance, the desire of the body to move, to be of use, and, as a child, to play. ADD is a decent alternative to simply going numb.

22. I think it is fair to assume that women know the nature of receptivity by virtue of the feminine body. Receptivity is embodied and reinforced in the act of making love. It becomes obvious (if the context has been well discerned) that receptivity is a good thing, that it is the ground of profound relationship.

23. I do not claim to know the phenomenology of the Beauty Way for traditional or contemporary Navajo (Dineh) people. I use the term "Beauty Way" because it so aptly describes life choices guided by the perception and creation of beauty.

4: Minding the Relations: Contrast, Qualities, and Patterns

1. D. Howes, 1991, p. 210.

2. Summarized in R. Greenway, 1993, p. 24.

3. P. Gold, 1994, p. 104.

4. Ibid., p. 105.

5. T. Roszak, 1982, p. 344.

6. In experiments involving a Ganzfeld, in which the visual field presented to the eye is entirely without contrast, the color of the field will fade to gray in a matter of seconds. This is true even when the field is backlit by specific wavelengths presented by a high-intensity monochromator.

7. J. Macy, 1991, p. 63.

8. D. Ackerman, in P. Murphy, 1993.

9. M. Riegner, in D. Seamon, ed., 1993.

10. G. Bateson, 1979, p. 8.

11. P. Devereux, 1992, p. 93.

12. L. E. Olds, 1992, p. 85.

13. L. Gray, 1995, p. 178.

14. F. Capra, 1996, p. 80.

15. G. Lakoff and M. Turner, 1989, p. xi.

16. D. Abram, 1996, p. 35.

17. H. Bortoft, 1996, pp. 73–74.

18. Ibid., p. 292.

19. Quoted in T. C. McLuhan, 1994, p. 20.

20. L. Van Der Post, 1957, p. 22.

21. J. Hillman, 1989, p. 23.

22. R. Johnson, 1986, p. 33.

23. J. Hillman, 1989, p. 23.

24. Quoted in R. Johnson, 1986, p. 27.

25. E. S. Casey, 1993, pp. 35–38.

26. D. Wood, 1999.

27. P. Gold, 1994.

5: Genuine Depth: Beyond Binocularity, Time, Space, and Fear

1. J. Baudrillard, 1979, pp. 58–59.

2. K. Wilber, 1997, p. 21.

3. D. Abram, 1988, p. 103.

4. E. S. Casey, 1993, p. 67.

5. J. Hillman, 1975, p. x.

6. G. Bateson, 1980, p. 252.

7. D. Meadows, 1991, p. 42.

8. J. Frueh, 1996, p. 7.

9. J. Hillman, 1989, p. 302.

6: Visionary Practice: Reversing the World

1. J. Spayde, 1995, p. 54.

2. P. Devereux, 1992. According to Devereux, this loosened ambiguity describes the common state of consciousness of the Neolithic human, of the "participating mind."

3. B. Plotkin, 1999.

4. P. B. Allen, 1995, p. 3.

5. J. Hillman, 1975, p. 23.

6. J. P. Sartre, 1996, p. 10.

7. Ibid., p. 31.

8. J. Hillman, 1989, p. 22.

9. J. Hillman, 1981, p. 7.

10. Ibid., p. 26.

11. S. M. Kosslyn, 1980.

12. M. Somé, 1994, p. 6.

13. R. F. Thompson, 1983, p. 5.

14. Quoted in T. C. McLuhan, 1994, p. 173.

15. According to an analysis done by the Wilderness Society, United States taxpayers lost $45 million on commercial logging programs in their national forests in fiscal year 1997 (*The Wilderness Society Quarterly Newsletter*, Spring 1999).

16. A. Zajonc, 1997, p. 30.

17. J. Hillman, 1975, p. 14.

18. P. Chodron, quoted in Chodron and A. Walker, 1999, p. 36.

19. E. Schiffmann, 1998.

7: Looking for a Worldview

1. K. Eagle Feather, 1995, p. 74.

2. D. Kidner, 1994, p. 359.

3. S. Hagen, 1995, p. 45.

4. K. Fackelmann, 1998, pp. 73–75.

5. J. Tuxill and C. Bright, in Brown et al., 1998, p. 52.

6. Ibid., p. 41.

7. Referenced in S. Ebenreck, 1996, p. 9.

8. L. Hyde, 1983, p. 193.

9. P. Gold, 1994, p. 53.

10. W. T. Anderson, 1997, p. 46.

11. A. Naess, 1988, p. 20.

12. D. Abram, 1996, p. 125.

13. P. Shepard, 1969, p. 112.

14. J. Hillman, 1995, p. xvii.

15. J. Elkins, 1996, p. 42.

16. Ibid., p. 44.

17. J. Hillman, 1989, pp. 18–19.

18. P. Rezendes, 1992, p. 23.

19. A. Fisher, unpublished manuscript, p. 152–53.

20. R. Nelson, 1989, p. 249.

21. J. Elkins, 1996, p. 43.

22. J. Hillman, 1993, p. 35.

23. From an all-time high of 2.2 percent the population growth rate began declining in 1967. In 1997, the rate was 1.4 percent (Brown et al., 1999).

24. A. Dillard, 1974, p. 201.

Selected Bibliography

Abraham, Ralph. *Chaos, Gaia, Eros: A Chaos Pioneer Uncovers the Three Great Streams of History.* San Francisco: HarperSanFrancisco, 1994.

Abraham, Ralph, Terence McKenna, and Rupert Sheldrake. *Trialogues at the Edge of the West.* Sante Fe: Bear and Company, 1992.

Abram, David. *The Spell of the Sensuous: Perception and Language in a More-Than-Human World.* New York: Pantheon Books, 1996.

Abram, David. "Merleau-Ponty and the Voice of the Earth." *Environmental Ethics,* Vol. 10 (Summer 1988), 101–120.

Ackerman, Diane. *A Natural History of the Senses.* New York: Vintage Books, 1990.

Allen, Pat B. *Art Is a Way of Knowing.* Boston: Shambhala, 1995.

Anderson, W. T. *The Future of the Self: Exploring the Post-Identity Society.* New York: Jeremy P. Tarcher/Putnam, 1997.

Arnheim, Rudolf. *Toward a Psychology of Perception: Collected Essays.* Berkeley: University of California Press, 1972.

———. *Art and Visual Perception: A Psychology of the Creative Eye.* University of California Press, 1983.

Arrien, Angeles. *The Four-Fold Way: Walking the Paths of the Warrior, Teacher, Healer and Visionary.* San Francisco: HarperSanFrancisco, 1993.

Bakan, Paul, ed. *Attention.* Princeton, NJ. Van Nostrand Company, Inc., 1966.

Barlow, H. B. and J. D. Mollon. *The Senses.* Cambridge: Cambridge University Press, 1982.

Bates, William. *The Bates Method for Better Eyesight Without Glasses*. Revised ed. New York: Henry Holt, 1986.

Bateson, Gregory. *Mind and Nature: A Necessary Unity*. New York: Bantam Books, 1979.

Baudrillard, Jean. *Seduction*. New York: St. Martin's Press, 1979.

Beebe, John. *Integrity in Depth*. New York: Fromm International Publishing Corporation, 1995.

Berger, John. *The Sense of Sight*. New York: Pantheon Books, 1985.

———. *Ways of Seeing*. London: British Broadcasting and Penguin Books, 1972.

Berman, Morris. *The Reenchantment of the World*. 2nd ed. New York: Bantam Books, 1988.

———. *Coming to Our Senses: Body and Spirit in the Hidden History of the West*. New York: Bantam Books, 1989.

Berry, Thomas. *The Dream of the Earth*. San Francisco: Sierra Club Books, 1988.

Birren, Faber. *Color Psychology and Color Therapy*. 3rd ed. Secaucus, NJ: The Citadel Press, 1980.

Blair, Lawrence. *Rhythms of Vision*. New York: Warner Books, 1975.

Bolles, Edmund Blair. *A Second Way of Knowing: The Riddle of Human Perception*. New York: Prentice Hall Press, 1991.

Bonaventure, Saint (Stephen F. Brown, ed., Philotheus Boehner, trans.). *The Journey of the Mind to God*. Cambridge: Hackett Publishing Company, 1956.

Boring, Edwin G. *A History of Experimental Psychology*. New York: Appleton-Century-Crofts, 1929.

Bortoft, Henri. *The Wholeness of Nature: Goethe's Way toward a Science of Conscious Participation in Nature*. New York: Lindisfarne Press, 1996.

Bower, Bruce. "All Fired Up: Perception May Dance to the Beat of Collective Neuronal Rhythms." *Science News*, Vol. 153 (February 1998), 120–21.

Breggin, Peter R., and Ginger Ross Breggin. *Talking Back to Prozac: What Doctors Won't Tell You About Today's Most Controversial Drug*. New York: St. Martin's Press, 1994.

Brennan, Teresa, and Martin Jay. *Vision in Context: Historical and Contemporary Perspectives on Sight*. New York: Routledge, 1996.

Brooks, Charles V. W. *Sensory Awareness: The Rediscovery of Experience through Workshops with Charlotte Selver*. 2nd ed. Great Neck, NY: Felix Morrow Publishers, 1974.

Brown, Lester, Christopher Flavin and Hilary French, eds. *State of the World, 1998*. New York: W. W. Norton and Company, 1998.

Brown, Lester, Christopher Flavin and Hilary French, eds. *State of the World, 1999.* New York: W. W. Norton and Company, 1999.

Brown, Jay, and Rebecca Novick. *Mavericks of the Mind: Conversations for the New Millennium.* Freedom: Crossland Press, 1993.

Brown, Tom. *Tom Brown's Field Guide to Nature Observation and Tracking.* New York: Berkeley Books, 1983.

———. *The Vision.* New York. Berkeley Books, 1988.

Bruce, Vicki, and Patrick R. Green. *Visual Perception: Physiology, Psychology and Ecology.* Hillsdale, IL: Lawrence Erlbaum Associates, 1990.

Cajete, Gregory. *Look to the Mountain: An Ecology of Indigenous Education.* Durango, CO: Kivaki Press, 1994.

Capra, Fritjof. *Uncommon Wisdom: Conversations with Remarkable People.* London: Bantam Books, 1988.

———. *The Web of Life: A New Scientific Understanding of Living Systems.* New York: Doubleday, 1996.

Casey, Edward S. *Getting Back into Place: Toward a Renewed Understanding of the Place-World.* Bloomington: Indiana University Press, 1993.

Chodron, Pema and Alice Walker. "Good Medicine for this World." *Shambhala Sun,* Vol. 7, No. 3 (January 1999), 32–61.

Cleary, Thomas. *Transmission of Light: Zen in the Art of Enlightenment.* San Francisco: North Point Press, 1990.

Clothiaux, Eugene E., Mark F. Bear and Leon N. Cooper. "Synaptic Plasticity in Visual Cortex: Comparison of Theory with Experiment." *Journal of Neurophysiology,* Vol. 66, No. 5 (1991), 1785–804.

Collier, Michael. *An Introduction to Grand Canyon Geology.* Grand Canyon National History Association, 1980.

Corbett, Jim. *Goatwalking: A Quest for the Peaceable Kingdom.* Middlesex: Bantam Books, 1991.

Cornsweet, Tom N. *Visual Perception.* Orlando, FL: Harcourt Brace Jovanovich, 1970.

Cottlieb, Roger S., ed. *This Sacred Earth: Religion, Nature, Environment.* New York: Routledge, 1996.

Crick, Francis. "Function of the thalamic reticular complex. The searchlight hypothesis." *Proceedings of the National Academy of Science,* 81 (1984), 4586–90.

———. *The Astonishing Hypothesis: The Scientific Search for the Soul.* New York: Maxwell Macmillan International, 1994.

Csikszentmihalyi, Mihaly. *Flow: The Psychology of Optimal Experience.* New York: Harper Perennial, 1990.

———. *The Evolving Self: A Psychology for the Third Millennium.* San Francisco: HarperSanFrancisco, 1993.

Csikszentmihalyi, Mihaly, and Rick E. Robinson. *The Art of Seeing: An Interpretation of the Aesthetic Encounter.* Malibu, CA: The Getty Center for Education in the Arts, 1990.

Curtois, Flora. "The Door to Infinity." *Parabola,* Vol. XV, No. 2 (Summer 1990), 17–19.

Devereux, Paul. *Symbolic Landscapes: The Dreamtime Earth and Avebury's Open Secrets.* Great Britain. Gothic Image Publications, 1992.

———. *Re-Visioning the Earth: A Guide to Opening the Healing Channels Between Mind and Nature.* New York: Simon and Schuster, 1996.

DiCarlo, Russell E., ed. *Towards a New World View: Conversations at the Leading Edge.* Erie, PA: Epic Publishing, 1996.

Dillard, Annie. *Pilgrim at Tinker Creek.* New York: Harper and Row, 1974.

Donis, D. A. *A Primer of Visual Literacy.* Cambridge: M.I.T. Press. 1973.

Eagle Feather, Ken. *A Toltec Path.* Charlottesville, VA: Hampton Roads Publishing Co., 1995.

Ebenreck, Sara. "Opening Pandora's Box: Imagination's Role in Environmental Ethics." *Environmental Ethics,* Vol. 18 (Spring 1996) 3–18.

Eck, Diana L. *Darśan: Seeing the Divine Image in India.* 2nd ed. Chamsberg, PA: Anima Books, 1985.

Eliade, Mircea. *Images and Symbols: Studies in Religious Symbolism.* Princeton: Princeton University Press, 1991.

Elkins, James. *The Object Stares Back: On the Nature of Seeing.* New York: Harcourt Brace and Company, 1996.

Fackelmann, Kathleen. "Tuberculosis outbreak: An ancient killer strikes a new population," *Science News,* Vol. 153, No. 5 (1998), 73–75.

Fisher, Andy. "Madness on Earth: Laying the Ground for an Ecopsychology." Unpublished manuscript, York University, 1996.

Forman, Richard, and Michel Godron. *Landscape Ecology.* New York: John Wiley and Sons, 1986.

Foster, Hal, ed. *Vision and Visuality.* Seattle: Bat Press, 1988.

Fox, Warwick. *Toward a Transpersonal Ecology: Developing New Foundations for Environmentalism.* Boston: Shambhala, 1990.

Frueh, Joanna. *Erotic Faculties*. Berkeley: University of California Press, 1996.

Gallagher, Winifred. *The Power of Place: How our Surroundings Shape our Thoughts, Emotions and Actions*. New York: Harper Perennial, 1993.

Gibson, James J. *The Ecological Approach to Visual Perception*. Boston: Houghton Mifflin Company, 1979.

Glendinning, Chellis. *My Name is Chellis and I'm in Recovery from Western Civilization*. Boston: Shambhala, 1994.

Gold, Peter. *Navajo and Tibetan Sacred Wisdom: The Circle of the Spirit*. Rochester: Inner Traditions, 1994.

Goldsmith, Edward. *The Way: An Ecological Worldview*. Boston: Shambhala, 1993.

Goldstein, E. Bruce. *Sensation and Perception*. Belmont, CA: Wadsworth Publishing Company, 1980.

Goldstein, Joseph. *The Experience of Insight: A Simple and Direct Guide to Buddhist Meditation*. London: Shambala. 1983.

Goodrich, Janet. *Natural Vision Improvement*. Berkeley: Celestial Arts, 1985.

Gottlieb, Roger S., ed. *This Sacred Earth: Religion, Nature, Environment*. New York: Routledge, 1996.

Gray, Leslie. "Shamanic Counseling and Ecopsychology: An Interview with Leslie Gray." In *Ecopsychology: Restoring the Earth, Healing the Mind*, edited by Theodore Roszak, Mary Gomes and Allen Kanner, 172-182. San Francisco: Sierra Club Books, 1995, p. 178

Gregory, Richard L. *Eye and Brain: The Psychology of Seeing*. 2nd ed. New York: McGraw-Hill, 1973.

————. *Odd Perceptions*. London: Routledge, 1986.

Gregory, Richard L., John Harris, Priscilla Heard, and David Rose, eds. *The Artful Eye*. Oxford: Oxford University Press, 1995.

Greenway, Robert. *Notes in Search of an Ecopsychology*, unpublished manuscript, 1993.

Griffin, Susan. *The Eros of Everyday Life*. New York. Doubleday, 1995.

Grifio, Francesca, and Rosenthal, Joshua. *Biodiversity and Human Health*. Washington, D.C.: Island Press, 1997.

Grudin, Robert. *The Grace of Great Things: Creativity and Innovation*. New York: Ticknor and Fields, 1990.

Gunaratana, Bhante Henepola. "Mindfulness and Concentration." *Tricycle: The Buddhist Review*. Fall 1998, 28–32.

Hagen, Steve. "Just Seeing." *The Quest*, (Autumn 1995) 42–80.

Halifax, Joan. *Shamanic Voices: A Survey of Visionary Narratives*. New York: Viking Penguin, 1979.

Hall, Edward T. *The Hidden Dimension,* Garden City, NY: 1966.

Halpern, Daniel, ed. *On Nature: Nature, Landscape and Natural History.* North Point Press, San Francisco, 1987

Hanh, Thich Nhat.*The Heart of Understanding.* Berkeley: Parallax Press, 1988a.

————.*The Sun in My Heart.* Berkeley, CA: Parallax Press, 1988b.

Harman, Willis. *Global Mind Change: The New Age Revolution in the Way We Think.* New York: Warner Books, 1988.

Hayward, Jeremy. *A Guide to Sacred Shambhala Warriorship in Daily Life.* New York: Bantam Books, 1995.

Heidegger, Martin. *Being and Time.* New York: Harper and Row, 1962.

Hejmadi, Padma. "Dhyana: The Long, Pure Look." *Parabola,* Vol. XV, No. 2 (Summer 1990), 70–75.

Heyneman, Martha. *The Breathing Cathedral: Feeling Our Way Into A Living Cosmos.* San Fransisco: Sierra Club Books, 1993.

Highwater, Jamake. *The Primal Mind: Vision and Reality in Indian America.* New York: Meridian Books, 1981.

Hillman, James. *Re-Visioning Psychology.* New York: Harper and Row, 1975.

————. *The Thought of the Heart and the Soul of the World.* Dallas: Spring Publications, 1981.

————. *A Blue Fire.* New York: Harper Collins Publishers, 1989.

————. "The Practice of Beauty." *Resurgence,* No. 157 (1993), 34–37.

————. "A Psyche the Size of the Earth." In *Ecopsychology: Restoring the Earth, Healing the Mind,* Theodore Roszak, Mary Gomes, and Allen Kanner, eds. San Francisco: Sierra Club Books, 1995, xvii–xxiii.

Hillman, James, and Michael Ventura. *We've Had a Hundred Years of Psychotherapy and the World's Getting Worse.* San Francisco: HarperSanFrancisco, 1992.

Hiss, Tony. *The Experience of Place.* New York: Vintage Books, 1990.

Houston, Jean. *The Possible Human: A Course in Enhancing Your Physical, Mental, and Creative Abilities.* New York: G. P. Putnam's Sons. 1951.

Howes, David, ed. *The Varieties of Sensory Experience: A Source Book in the Anthropology of the Senses.* Toronto: University of Toronto Press, 1991.

Huxley, Aldous. *The Art of Seeing.* Seattle: Montana Books, 1942.

Huxley, Francis. *The Way of the Sacred.* London: Bloomsbury Books, 1989.

Hyde, Lewis. *The Gift: Imagination and the Erotic Life of Property.* New York: Vintage Books, 1983.

Jackson, John Brinckerhoff. *A Sense of Place, A Sense of Time.* New Haven: Yale University Press, 1994.

Jackson, Michael. *At Home in the World.* Durham: Duke University Press, 1995.

Jensen, Derrick. *Listening to the Land: Conversations about Nature, Culture, and Eros.* San Francisco: Sierra Club Books, 1995.

Johnson, Mark. *Moral Imagination.* Chicago: University of Chicago Press, 1993.

Johnson, Robert A. *InnerWork: Using Dreams and Active Imagination for Personal Growth.* San Francisco: Harper and Row, 1986.

Jung, Carl, ed. *Man and His Symbols.* New York: Dell Publishing, 1964.

Kahneman, D. *Attention and Effort.* Englewood Cliffs: Prentice-Hall, 1973.

Kasamatsu, T. "Adrenergic regulation of visuocortical plasticity: A Role of the Locus Coeruleus system." *Progress in Brain Research,* Vol. 88 (1991): 599–616.

Kawagley, A. Oscar. *A Yupiaq Worldview: A Pathway to Ecology and Spirit.* Prospect Heights, IL: Waveland Press, Inc., 1995.

Kaza, Stephanie. *The Attentive Heart: Conversations with Trees.* New York: Fawcett Columbine, 1993.

Kellert, Stephen, and Edward O. Wilson, eds. *The Biophilia Hypothesis.* Washington, DC: Island Press, 1993.

Kidner, David W. "Why Psychology is Mute about the Environmental Crisis." *Environmental Ethics,* Vol. 16, No. 4. (Winter 1994), 359–77.

Klocek, Dennis. *Seeking Spirit Vision: Essays on Developing Imagination.* Fair Oaks, CA: Rudolph Steiner Press, 1998.

Knopp, Lisa. *Field of Vision.* Iowa City: University of Iowa Press, 1996.

Kornfield, Jack. *A Path with Heart: A Guide through the Perils and Promises of Spiritual Life.* New York: Bantam Doubleday Dell, 1993.

Kosslyn, Stephen M. *Image and Mind.* Cambridge, MA: Harvard University Press, 1980.

———. "Aspects of a Cognitive Neuroscience of Mental Imagery." *Science,* Vol. 240 (June 1988), 1621–26.

LaChapelle, Dolores. *Sacred Land, Sacred Sex, Rapture of the Deep: Concerning Deep Ecology and Celebrating Life.* Silverton, CO: Finn Hill Arts, 1988.

Lakoff, George, and Mark Turner. *More than Cool Reason: A Field Guide to Poetic Metaphor.* Chicago: University of Chicago Press, 1989.

Langer, Ellen. *The Power of Mindful Learning.* Reading, MA: Addison-Wesley, 1997.

Langer, Monika M. *Merleau-Ponty's Phenomenology of Perception: A Guide and Commentary.* Tallahassee: Florida State University Press, 1989.

Lawlor, Robert. *Voices of the First Day: Awakening in the Aboriginal Dreamtime.* Rochester, NY: Inner Traditions, 1991.

Leonard, George, and Michael Murphy. *The Life We Are Given: A Long Term Program for Realizing the Potential of Body, Mind, Heart, and Soul.* New York: G. P. Putnam's Sons, 1995.

Lindberg, David C. *Theories of Vision from Al-Kindi to Kepler.* Chicago: University of Chicago Press, 1976.

Llinas, Rodolfo R., ed. *The Workings of the Brain: Development, Memory and Perception: Readings from Scientific American.* New York: W. H. Freeman and Company, 1989.

Lopez, Barry. *Arctic Dreams: Imagination and Desire in a Northern Landscape.* New York: Bantam Books, 1996.

―――. *Crossing Open Ground.* New York: Vintage Books, 1989.

Mack, Arien and Irvin Rock. *Inattentional Blindness.* Cambridge, MA: M.I.T. Press, 1998.

McLuhan, T. C. *The Way of the Earth: Encounters with Nature in Ancient and Contemporary Thought.* New York: Simon and Schuster, 1994.

McNamee, George, ed. *Named in Stone and Sand: An Arizona Anthology,* University of Arizona Press, Tucson, 1993.

Macy, Joanna. *Mutual Causality in Buddhism and General Systems Theory.* Albany, NY: State University of New York Press, 1991.

Mander, Jerry. *In the Absence of the Sacred: The Failure of Technology and the Survival of the Indian Nations.* San Francisco: Sierra Club Books, 1992.

Manes, Christopher. Untitled interview in Derrick Jensen, *Listening to the Land: Conversations about Nature, Culture and Eros.* San Francisco: Sierra Club Books, 1995.

Marcus, Morton. *Origins: Poems by Morton Marcus.* Santa Cruz, CA: Kayak Books, 1969.

Meadows, Donella. *The Global Citizen.* Washington, DC: Island Press, 1991.

Meadows, Donella, Dennis Meadows, Jorgen Randers, and William Behrens III. *The Limits of Growth: A Report for the Rome's Project on the Predicament of Mankind.* New York: Universe Books, 1974.

Meyers, Steven J. *On Seeing Nature.* Golden, CO: Fulcrum, 1987.

Miller, Izchak. *Husserl, Perception and Temporal Awareness.* London: M.I.T. Press, 1984.

Mookerjee, Ajit, and Rhanna Madhu. *The Tantric Way: Art, Science, Ritual.* Boston: New Geographic Society, 1977.

Moss, Richard. *The Black Butterfly: An Invitation to Radical Aliveness.* Berkeley: Celestial Arts Publishing, 1986.

―――. *The Second Miracle: Intimacy, Spirituality, and Conscious Relationships.* Berkeley: Celestial Arts Publishing, 1995.

Murphy, Pat, and William Neill. *By Nature's Design: An Exploratorium Book.* San Francisco: Chronicle Books, 1993.

Naess, A. "Self-Realization: An Ecological Approach to Being in the World." In John Seed, Joanna Macy, Pat Fleming, and Arne Naess, *Thinking Like a Mountain: Towards a Council of All Beings.* Philadelphia: New Society Publishers, 1988, 19–30.

Nagel, Thomas. *The View from Nowhere.* New York: Oxford University Press, 1986.

Nelson, Richard. *The Island Within.* New York: Vintage Books, 1996.

Novak, Philip. "The Practice of Attention." *Parabola*, Vol. XV, No. 2 (Summer 1990), 5–13.

Nunn, Chris. *Awareness: What It Is, What It Does.* London: Routledge, 1996.

Olds, Linda E. *Metaphors of Interrelatedness: Toward a Systems Theory of Psychology.* Albany, NY: State University of New York Press, 1992.

Ornstein, Robert, and Paul Ehrlich. *New World New Mind: A Brilliantly Original Guide to Changing the Way We Think About the Future.* New York: Simon and Schuster, 1989.

Orr, David W. *Earth in Mind: On Education, Environment and the Human Prospect.* Washington, DC: Island Press, 1994.

Ortiz, Alfonso. *The Tewa World: Space, Time, Being and Becoming in a Pueblo Society.* Chicago: University of Chicago Press, 1969.

Ott, John. *Health and Light: The Effects of Natural and Artificial Light on Man and other Living Things.* Columbus, OH: Ariel Press, 1973.

Pastore, Nicholas. *Selective History of Theories of Visual Perception: 1650–1950.* New York: Oxford University Press, 1971.

Perkins, David N. *The Intelligent Eye: Learning to Think by Looking at Art.* Santa Monica, CA: The Getty Center for Education and the Arts, 1994.

Plotkin, Bill. *SoulCraft: Nature, Self and Society,* unpublished manuscript, 1999.

Pyle, Robert Michael. *The Thunder Tree: Lessons from an Urban Wildland.* Boston: Houghton Mifflin, 1993.

Rauschecker, J. P. "Mechanisms of Visual Plasticity: Hebb Synapses. NMDA Receptors and Beyond." *Physiology Review,* Vol. 71, No. 2 (1991): 587–615.

Riegner, Mark. "Toward a Holistic Understanding of Place: Reading a Landscape Through Its Flora and Fauna," in Seamon, David, ed., *Dwelling, Seeing, and Designing: Toward a Phenomenological Ecology.* Albany, NY: State University of New York Press, 1993.

Rezendes, Paul. *Tracking and the Art of Seeing: How to Read Animal Tracks and Sign.* Charlotte, VT: Camden House Publishing, 1992.

Rock, Irvin, ed. *The Perceptual World: Readings from* Scientific American. New York: W. H. Freeman and Company, 1990.

———. *Perception.* New York: Scientific American Books, Inc., 1995.

Rossbach, Sarah, and Lin Yun. *Living Color: Master Lin Yun's Guide to Feng Shui and the Art of Color.* New York: Kodansha International, 1994.

Roszak, Theodore. *Where the Wasteland Ends: Politics and Transcendence in Post Industrial Society.* Berkeley: Celestial Arts Publishing, 1982.

———. *The Voice of the Earth.* New York: Simon and Schuster, 1992.

Roszak, Theodore, Mary E. Gomes, and Allen Kanner. *Ecopsychology: Restoring the Earth, Healing the Mind.* San Francisco: Sierra Club Books, 1995.

Samuels, Mike, and Nancy Samuels. *Seeing with the Mind's Eye.* New York: Random House, 1975.

Sartre, Jean Paul. *The Psychology of Imagination.* New York: Washington Square Press, 1966.

Seamon, David, ed. *Dwelling, Seeing, and Designing: Toward a Phenomenological Ecology.* Albany, NY: State University of New York Press, 1993.

Schiffmann, Erich. "Moving into Stillness." *Yoga Journal,* No. 143 (Nov/Dec 1998): 52–66.

Shepard, Paul. *Traces of an Omnivore,* Washington, D.C.: Island Press, 1996.

———. *The Others: How Animals Made Us Human.* Washington, DC: Island Press, 1996.

Sheppard, Roger N. *Mind Sights: Original Visual Illusions, Ambiguities, and Other Anomalies.* New York: W. H. Freeman and Company, 1990.

Simmons, I. G. *Interpreting Nature: Cultural Constructions of the Environment.* New York: Routledge, 1993.

Sinclair, Sandra. *How Animals See: Other Visions of Our World.* New York: Facts on File Publication, 1985.

Singer, Wolf. "Development and Plasticity of Cortical Processing Architectures." *Science,* Vol. 270 (5237) (1995): 758–764.

Schwenk, Theodore. *Sensitive Chaos; The Creation of Flowing Forms in Water and Air.* New York: Schocken Books, 1976.

Smith, Huston. *The World's Religions.* San Francisco: HarperSanFrancisco, 1992.

Snyder, Gary. *The Practice of the Wild.* San Francisco: North Point Press, 1990.

———. *A Place in Space.* Washington DC: Counterpoint, 1995.

Somé, Malidoma Patrice. *Of Water and the Spirit: Ritual, Magic and Initiation in the Life of an African Shaman*. New York: Penguin Books, 1994.

Sommer, Robert. *The Mind's Eye: Imagery in Everyday Life*. New York: Dell Publishing, 1978.

Spayde, Jon. "One Hundred Visionaries that Could Change Your Life." *Utne Reader*, No. 67 (1995): 54–81.

Spillman, Lothar, and John S. Werner. *Visual Perception: The Neurophysiological Foundations*. New York: Academic Press/Harcourt Brace Jovanovich, 1990.

Spoehr, Kathryn, and Stephen Lehmkuhle. *Visual Information Processing*. San Francisco: W. H. Freeman and Company, 1982.

Staniswzewski, Mary Anne. *Believing Is Seeing: Creating the Culture of Art*. New York: Penguin Books, 1995.

Steinberg, David, ed. *The Erotic Impulse: Honoring the Sensual Self*. New York: Jeremy Tarcher, 1992.

Suzuki, D. T. *The Lankavatara Sutra: A Mahayana Text*. Taipei: SMC Publishing, 1991.

Tarnas, Richard. *The Passion of the Western Mind: Understanding the Ideas that Have Shaped Our World View*. New York: Harmony Books, 1991.

Thomashow, Mitch. *Ecological Identity: Becoming a Reflective Environmentalist*, Cambridge, MA: M.I.T. Press, 1995.

———. "Very, Very Deep Ecology: Three Books by Lynn Margulis and Dorion Sagan." *Orion*, Vol. 17, No. 1 (1998): 68–69.

Thompson, Robert Farris. *Flash of the Spirit: African and Afro-American Art and Philosophy*. New York: Vintage Books, 1984.

Trimble, Stephen, ed. *Words from the Land*. Salt Lake City: Peregrine Smith Books, 1988.

Tuan, Yi-Fu. *Topophilia: A Study of Environmental Perception, Attitudes, and Values*. New York: Columbia University Press, 1974.

Tuchman, Barbara W. *A Distant Mirror: The Calamitous 14th Century*. New York: Alfred A. Knopf, 1978.

Turner, Frederick. *Rebirth of Value: Meditations on Beauty, Ecology, Religion, and Education*. Albany, NY: State University of New York Press, 1991.

Van Der Post, Laurens. "The Creative Pattern in Primitive Africa." Ascona, Switzerland: Eranos Foundation, 1957.

Varela, Francisco, Evan Thompson, and Eleanor Rosch. *The Embodied Mind: Cognitive Science and Human Experience*. Cambridge, MA: M.I.T. Press, 1997.

Vickery, Dale. *Wilderness Visionaries*. Minocqua, WI: Northward Press, 1986.

Volk, Tyler. *Metapatterns: Across Space, Time, and Mind*. New York: Columbia University Press, 1995.

Wallace, Alan. "Training the attention and exploring consciousness in Tibetan Buddhism." *Consciousness Bulletin*, Fall 1998: 5–7.

Wessels, Tom. *Reading the Forested Landscape: A Natural History of New England*. Woodstock, NY: The Countryman Press, 1997.

White, Jonathan. *Talking on the Water: Conversations about Nature and Creativity*. San Francisco: Sierra Club Publishing, 1994.

Wilbur, Ken. *Eye to Eye: The Quest for the New Paradigm*. Boston: Shambhala, 1990.

———. *A Brief History of Everything*. Boston: Shambhala, 1996.

———. *The Eye of the Spirit: An Integral Vision for a World Gone Slightly Mad*. Boston: Shambhala, 1997.

Williams, Terry Tempest. *An Unspoken Hunger: Stories from the Field*. New York: Vintage Books, 1994.

Wood, Daniel, "Caught in a Tangled Web of US-Indian History." *Christian Science Monitor*, Jan. 26, 1999: 12–13.

Zajonc, Arthur. *Catching the Light: The Entwined History of Light and Mind*. New York: Bantam Books, 1993.

———. "Toward a Yoga of the Senses: An Interview with Arthur Zajonc." *Orion*, Vol. 16, No. 4 (1997): 26–31.

Index